THE SCEPTICAL BOTANIST

SEPARATING Fact from Fiction

TIM ENTWISLE

PUBLISHING

A catalogue record for this book is available from the National Library of Australia

ISBN: 9781486318216 (pbk)
ISBN: 9781486318223 (epdf)
ISBN: 9781486318230 (epub)

Published by:

CSIRO Publishing
36 Gardiner Road, Clayton VIC 3168
Private Bag 10, Clayton South VIC 3169
Australia

Telephone: +61 3 9545 8400
Email: publishing.sales@csiro.au
Website: www.publish.csiro.au
Sign up to our email alerts: publish.csiro.au/earlyalert

Front cover: Wild flowers, banksia flower, eucalyptus flowers and bee (graphics by Yumeee, Tommy Atthi, Sketchbook Designs and Shanvood, respectively, via Shutterstock.com)
Spine: Grevillea (graphic by Sketchbook Designs via Shutterstock.com)
Back cover: Golden yellow flowers and wild flowers (graphics by Sketchbook Designs and Yumeee via Shutterstock.com)

Illustrations are by Jerome KS Entwisle

Edited by Janet Walker
Cover design by Cath Pirret
Typeset by Envisage Information Technology
Printed by Ingram Lightning Source

CSIRO acknowledges the Traditional Owners of the lands that we live and work on across Australia and pays its respect to Elders past and present. CSIRO recognises that Aboriginal and Torres Strait Islander peoples have made and will continue to make extraordinary contributions to all aspects of Australian life including culture, economy and science. CSIRO is committed to reconciliation and demonstrating respect for Indigenous knowledge and science. The use of Western science in this publication should not be interpreted as diminishing the knowledge of plants, animals and environment from Indigenous ecological knowledge systems.

Oct25_RP_ILS

CONTENTS

PREFACE

Why?

I am a scientist, by training and inclination, so being sceptical and (constructively) argumentative is how I am. That said, moving through life and my career in management, I've found that moderating my inclination to argue has been productive and well received. Thankfully, I remain stubbornly sceptical about implausible and outrageous assertions: as scientist and author Carl Sagan reminded us, extraordinary claims need extraordinary evidence.

The downside to such a critical approach to life, albeit moderated by experience, is the risk of missing unlikely and unexpected truths. To counter that tendency, I try to admit an unlikely idea *if* the proponent is willing to expose it to scientific scrutiny or at least admit the possibility that it may be false. But that is a big and important *if*.

Plant 'intelligence', the first topic I cover in this book, is a good example. I'm as sceptical as hell about traits ascribed to plants that somehow bequeath them with consciousness. On the other hand, why not? If plants are conscious, so what? That would be an unexpected but interesting truth.

Same with planting only when the Moon has a particular form in our night sky. This seems an outrageous idea. Yet who knows, maybe there is a relationship between the Moon and your tomato plant that may turn out to be true. It is just that we haven't yet found proof it is so.

And I'm not just being 'clever'. I acknowledge that life is a funny old thing and my take on it is just one in many billion possible views. These essays are my attempt to make sense of things, initially to satisfy my own curiosity. Sometimes, you, my reader, will cheer. Other times I expect you'll disapprove, strongly.

My conclusions on genetically modified organisms, glyphosate use and organic gardening will not please everyone. Nor will my permissive approach to plants from other places. I've always been intrigued by how differently we all react to a plant once it has been labelled a 'weed' or 'invasive'. A plant can shift from being beloved to despised in one simplistic nomenclatural manoeuvre. It's almost magical.

So, from stupid plants to misguided humans, I hope my sceptical deliberations will entertain, provoke and inform. How you deal with these truths and possible truths is entirely up to you, but my son Jerome provides his responses in pen and ink.[1]

About me

I began writing about science for a general audience in late 1989, towards the end of postdoctoral fellowship at the University of Melbourne. Within hours of attending a talk by the science editor for *The Age*, Graeme O'Neill, I was writing the first of what became an almost monthly contribution to that newspaper. After a year or so, *The Age* reduced both its coverage of science and the desire to employ freelance writers, so I concentrated on my day job and became a phycologist, botanist and then botanic garden director.

Not entirely though. I continued to write in the early morning and evenings, or when I had an hour or two spare on weekends. For a while I was a regular feature writer for *Nature Australia*, then later and continuing today, a regular contributor to *Gardening Australia* magazine. Every now and then, I fired off an opinion piece to the local newspapers in Sydney, London or Melbourne – depending on where I was living. Some of these essays are those, first published in *Sydney Morning Herald*, *The Australian*, *Telegraph*, *The Guardian* or *The Age*.

While this is my first compilation of essays, new or recycled, to some extent all my books have been written as a collection of short pieces, initially for technical reasons, given that most of them are compilations of scientific information.

In the late 1990s, I co-edited and co-authored the extensive (and expensive) *Flora of Victoria*,[2] a guide to the vascular plants growing naturally (and naturalised) in Victoria. Around the same time, I was lead author of *Freshwater Algae in Australia*,[3] the first colour field guide to Australia's inland algae. Then in 2014, the Commonwealth Scientific and Industrial Research Organisation (CSIRO) published *Sprinter and Sprummer: Australia's Changing Seasons*,[4] a collection of my observations on the appropriateness of the four seasons in Australia and my proposal for a fit-for-purpose system.

Most recently, Thames and Hudson commissioned me to write a memoir on my time in botanic gardens. Published in 2022, *Evergreen: The Botanical Life of a Plant Punk*[5] is resplendently decorated with subheadings, a wayfinding device for the author as much as the reader. A book, like a life, has its waypoints.

About the essays

Around half the essays in this book were first published elsewhere, mostly in my bimonthly column 'Curiosities' in *Gardening Australia* magazine (and only available in print form). A few draw heavily on published work but have been extensively expanded or updated in this book. Together, the essays cover a period of 25 years, with the first published by *The Age* in 1989.

At the end of each essay, I've provided the source for any of my previously published work, although in most cases they have been edited lightly to improve grammar and readability (to my current standards and biases) and to minimise duplication with other essays. Sometimes I've excised a paragraph or two that seem incongruous or gratuitous today. In a few cases, I've combined different versions of an essay or incorporated into a new one some previously published observations; where these previously published contributions are significant, I've acknowledged this.

A few of the new essays started life as posts in my long-running blog, *Talking Plants*,[6] or as notes for the ABC RN radio show of that title I hosted[7] and my regular contributions to ABC RN *Blueprint for Living*,[8] but all have been substantially rewritten and expanded from their original form. I provide a link to the source again only where the debt to the original is, in my mind, significant.

While the essays are not referenced in full like a peer-reviewed scientific paper, I've provided some sources where I feel a statement begs additional support or where I've incurred a major debt to another author. Generally, though, a thorough (and sceptical) trawl of internet sources or a good library will provide you with the same sources I had.

Finally, I've added a postscript to all essays not prepared new for this book so that I can flag any major advance in knowledge since they were written, reflections on how they were received, or where my views have changed since. Sometimes there is little I need to say.

1

What makes plants tick?

My botanical origin myth has me entering university as a maths/physic nerd and emerging a botanist. In truth, I emerged a phycologist – an algal nerd – but I transitioned through botany to get there. My first year curriculum was packed with advanced physics and 'pure' maths (I wasn't keen on the 'applied' kind), leaving room for a discretionary subject. I chose botany because I was curious about the natural world but didn't want to cut up animals.

What eventually converted me to the plant (and algal) world – as I'm fond of saying – was the image of a plant cell projected onto a wall in the Old Botany Theatre at the University of Melbourne. Within the cellulose impregnated cell wall there were sacks of DNA, to run the show, and others crammed with photoreceptors and the apparatus needed for the solar-powered conversion of carbon dioxide into oxygen and sugar. Along with other bibs and bobs. Who would have known?

A teacher friend took me aside once after I told this story and answered this question for me. Well, she said, 'You, for a start.' Despite not having studied biology at secondary school, it was apparently inconceivable I hadn't been told in junior science that plants were made up of cells containing things like nuclei and chloroplasts.

In any case, the image did the trick. I dropped maths and physics and headed full throttle into botany, seasoned with a little organic chemistry and genetics to help me understand what went on inside those cells.

By inclination, then, I seem to be a reductionist. To understand, I first break something down into its constituent parts, then reassemble as each part becomes knowable. Add to that my acquired trade as a taxonomist, where I look for characters – shared and unique – to make sense of the world created by those plant cells, and you begin to understand why detail matters to me.

To be honest, though, an encounter with an unknown plant or alga typically begins with its *gestalt*, and then moves on to a more considered assessment through reducing it to a collection of traits. I apply the same approach in this first selection of essays, starting, I'm sure, with pre-formed views of some kind and then picking away at the question until I confirm my bias or – and yes, it does happen, good scientist that I am – I persuade myself to change my mind.

Some of these phenomena are well understood but often misconceived. They include why we don't water our gardens in the middle of day, why some hydrangeas are pink, and why (almost) no amount of plant and flower material in your bedroom will suffocate you.

Others present me with that inconvenient or uncomfortable truth – that we just don't know; an observation that cannot be easily explained, or one that is no better understood through reductionism. This includes plants that seem intelligent, plants that 'talk' and why patting a plant might be good for it. In time I expect these extraordinary ideas to be proved or disproved, but for now, they mock me like the smile of Lewis Carroll's Cheshire Cat,[1] who, it must be said, was often right.

Stupid plants

Plants don't need to be like us to be smart

A lot has been written in recent years about plants being smart. Believe it all and you'll suffer nauseating guilt and regret every time you eat a carrot. You certainly wouldn't hold a Cabinet meeting near a scribbly gum or trumpet vine. At best, you might seek to commune with the green sentient beans in your vegetable garden.

It's frenetic out there. Drawing on chemical stockpiles the envy of *Breaking Bad*'s Walter White, plants can fend off hostile insect attacks by calling in squadrons of predatory wasps, at the same time warning their vegetable cohorts to prepare arms. Peas have been overheard conversing in clicks, reminiscent of the Khoisan in South Africa.

I'm sure you know by now that under our very feet, kilometres of fungal threads connect forest trees into a real, rather than Middle, Earth version of J.R.R. Tolkien's Fangorn forest, populated not by phlegmatic Ents but by collaborative beech and oak.[2] Plants (and fungi) do all this apparent thinking without the need for that distracting mush we carry around inside our heavy, bony skulls.

There is some evidence for these 'behaviours' and some justification for the view that we tend to underestimate the ability of plants to respond (rather sensibly in general) to the world around them. When this argument ratchets up to plants having some kind of neurology, or capacity to think in a way that we might equate with humans or other animals, I start to feel emotionally and intellectually uncomfortable.

Here is where I stand. Plants are fascinating: they do things animals don't do (such as convert carbon dioxide into sugars and oxygen using solar power) and they live at a pace almost unconceivable to humans. Slowly. They move, respond to the environment, possibly make relatively complex decisions and undoubtedly do a whole lot of things we don't yet understand, but mostly at a (to us) relaxed pace. For botanists or plant scientists they are always going to be more interesting than animals. Then people go and ruin it all by wanting plants to be intelligent or do intelligent things like we do.

Years ago, I was quite taken by *In Praise of Plants*,[3] a charming book written by cheeky French botanist Francis Hallé. He argues eloquently, and at times rather perversely, for the elevation of plants above animals in almost every respect, and to end our indifference to the vegetable world. He also says plants do some apparently very smart things, like respond to external stimuli in consistent and clever ways. Such responses, he says, are programmed and therefore always correct, making plants 'not intelligent because unlike humans they do not make mistakes'.

I can live with that as a defining statement about plant intelligence. I'm less relaxed about recent attempts – even as metaphors – to equate plant responses to an intelligence one might find in a sentient animal. While driven by an admirable intent (plants really are interesting), ultimately this kind of reasoning seems at best to turn our flora into simply slow fauna.

What defines this slow fauna movement? In his widely read 2013 *New Yorker* article[4] on the intelligent plant, Michael Pollan begins by referencing *The Secret Life of Plants* by Peter Tompkins and Christopher Bird. Published in 1973, it was a book that I bought second-hand a decade or so after publication, chuckling at its breathless reporting of Soviet insights into plant intelligence and signs of intelligent life in our vegetable companions. I remember, as Pollan does, Tompkins and Bird's excitement at the experiments of Cleve Backster, who found that plants remembered things, reacted morally and experienced stress, all pretty much like we do.

Much of the science in that book has been discredited but as Pollan notes, some, such as Daniel Chamovitz (author of *What a Plant Knows*),[5] consider the book still damaging to 'the cause', because scientists 'became wary of any studies that hinted at parallels between animal senses and plant senses'. In 2005, Pollan reports, there was a push for a new field of 'plant neurobiology' by some plant scientists, heavily resisted by other scientists as lacking any evidence. It isn't that plants don't use chemical and electrical signalling, say the sceptics (such as me), just that the mechanisms used are quite unlike those of an animal nervous system.

Pollan cites Lincoln Taiz, a plant physiologist from University of California Santa Cruz, who is particularly scathing about the plant intelligence concept. The plant neurobiologists, says Taiz, suffer from 'over-interpretation of data, teleology, anthropomorphizing, philosophizing, and wild speculation'. All will be explained by chemical or electrical pathways, he says.

According to Stefano Mancuso, defender of plant intelligence, plants do it differently to animals, and that stops us from truly appreciating the way they work. Because most are rooted in the ground, they develop responses to what life brings to them rather than new ways to flee. Hence their modular design,[6] which creates resilience and redundancy. The lack of irreplaceable organs in many plants means they can survive grazing and other damage that would be catastrophic to most animals.

In particular, says Mancuso, plants have a 'highly developed sensory apparatus', amounting to 15–20 distinct senses. This includes 'hearing', demonstrated by a plant's ability to detect caterpillars munching and water flowing through nearby soil. Forester and author Peter Wohllenben – who I'll come to in a moment – claims that because trees respond differently to the saliva of different insects, 'they must also have a sense of taste'.

Pollan's essay features controversial Perth scientist, Dr Monica Gagliano. I have interviewed Gagliano on this subject,[7] fascinated then and now by anyone who is willing to question our assumptions and use science to test what seem to be unlikely realities. At the scientific meeting Pollan attended, Dr Gagliano's theories about her recent research received a cool reception.

The plant under study was a mimosa, often called the 'sensitive' plant, because its fern-like leaves fold inwards when the wind blows or you touch them. Dr Gagliano found that if you drop the plant (in a pot) repeatedly, the leaf responds at first as it does when you touch it, but after a while remains open despite the jolting.

This response was interpreted as 'remembering', with the plant not closing its leaves to a similar response weeks later. These experiments were dismissed by some in the audience as being too contrived, confusing tissue fatigue with habituation and mixing adaptation with learning.

The criticism of Gagliano's work was in part around her use of language rather than the results of the experiments themselves. There was, for many, too much anthropomorphism.

Pollan ends as I would. Plants are fascinating, but in an entirely different way to animals. That is, we don't need to comfort ourselves by using complex terms like 'plant neurobiology' but rather celebrate things that plants do, like photosynthesis, and (with caveats)[8] the 'cooperative underground economy' of the forest, the wood-wide web. No animal can compete with that, acknowledging that fungi – neither plant nor animal – are a critical part of that purported subterranean web.

Which brings me nicely to the most recent treatise on this topic, *The Hidden Life of Trees: What they Feel, How they Communicate, Discoveries from a Secret World* by Peter Wohlleben.[9] This book makes me uneasy, if not a little queasy. I love the passion and intent but not the misty-eyed world view and the inflating of a few facts into grand theories. Wohllenben starts with a tree stump, which without leaves to provide it with fresh food remains alive because it is connected by cooperative or co-opted fungal threads to the roots of nearby leafy trees (yes, the wood-wide web again).

Wohllenben argues all trees are naturally interdependent in a very conscious way. They exchange nutrients and 'help neighbours', so that a forest is a 'superorganism with interconnections much like ant colonies'. But more than this, trees are, apparently, 'reluctant to abandon their dead', 'look after their own' and 'help their sick and weak back up onto their feet'. Trees do play favourites, though, and Wohllenben wonders why some are 'kept alive over the centuries' and others are not.

Wohllenben is impressed, as he should be, by the smart things plants do. Beech trees emit electrical signals in response to munching caterpillars, which in 'an hour or so' trigger the release of toxic chemicals from leaves in other parts of the plant. African acacias[10] respond to giraffe grazing by producing toxins in their leaves as well as a gas, ethylene, which causes nearby trees of the same species to release the same toxins – giraffes apparently stop grazing when they taste the toxins, and move on not to the next tree but to one 100 metres or so away.

This is all very interesting until Wohllenben treats plants as sentient beings. In addition to the examples above about plants helping each other, disturbances to forests are 'traumatic', injuries to plants are 'painful', stresses must be 'borne' and trees 'die prematurely'. He also adds a dubious moral overlay about what constitutes 'nature'.[11] Selective breeding has stripped *cultivated* plants of their ability to communicate, leaving them 'easy prey for insect pests'. Giant redwoods planted in Europe don't grow as tall as those in their native habitat in California because they are *not planted in a forest* with members of their own species.[12]

Things get even weirder when Wohllenben claims that trimming the roots of saplings, as would be done in a commercial nursery, risks the loss of 'brain-like structures ... along with the sensitive tips'.

Alongside other writers, including (obliquely) Charles Darwin in his and Francis Darwin's book *Power of Movement in Plants*, Wohlleben punts for a plant 'brain' existing in the plant's roots, and particularly the root tips. Presumably the network of roots and fungal hyphae conjure up an image of the human brain with its neuron connections. A root also 'feels' its way through the soil and negotiates rocks and unsuitable substrate. So somehow, when a tree communicates with insects and other pollinators using perfumes in their flowers, or through various visual stimuli, this is orchestrated from the root tips, I assume.

Anyway, putting the location of the plant brain to one side, the selfless cooperation between individuals of a species also stretches my credulity. 'Trees synchronize their performance so that they are all equally successful', which is why according to Wohlleben trees will never grow *too* close to each other. A tree is 'only as strong as the forest that surrounds it'. In a forest, are the vibrations caused by interruptions to the flow of water in roots 'cries of thirst'? Perhaps 'dire warnings to their colleagues that water levels are running low'. And so on, and so on, he writes.

While for Francis Hallé and Michael Pollen it's *vive la difference*, Wohlleben thinks we get obsessed with trying to keep some boundary between animals and plants, hence our reluctance to ascribe intelligence of any kind to vegetable matter. In his mind, more attention should be paid to the similarities between plants and animals. In my mind, we should praise our plants and celebrate their peculiarities. We love them not as conscious beings but because life on Earth depends on them and because for many of us humans, they are what makes life worth living.

The Skeptic, March 2017[13]

Postscript

During my time as Director of Sydney's botanic gardens in the early 2000s, I prepared a set of presentations that I called (privately) my

Bizarre Plant Lectures. The first of these, 'What plants do after dark', I delivered for the first time to a dozen or so weary shoppers at David Jones in a lunchtime entertainment event leading up to Christmas. The two follow-up lectures, 'Towering trees and flamboyant flowers' or 'Triffids – plants that eat, kill and move', didn't get an outing that year.

My personal favourite from the set was 'Are plants immortal and do they care?', which leaned heavily on a few humorous sketches from Hallé's *In Praise of Plants*, leavened further with a couple of the more peculiar assertions from *The Secret Life of Plants*. This became my public lecture staple for many years, culminating in an after-dinner delivery to 42 New South Wales Supreme Court judges in 2010.

It was through writing and delivering this presentation that I coalesced my views on what has become a rather tiresome topic: the intelligence of plants. My conclusion, as you have just read, is that they do clever things – many of them things we can't do – but to rank them against human intelligence or sentience is a futile and, I think, empty pursuit.

That said, I remain open-minded and curious about how plants interact with the world. I have no doubt there are some surprising and unsettling discoveries to be made. I don't expect these responses to be of the scale and nature asserted by Monica Gagliano, but I follow her and her colleagues research with interest.

Heard it on the grapevine

Are plants 'literally' crying out for help?

Just when you thought it was safe to return to the garden, scientists warn us of plants screaming, complaining or, on a quiet day, chattering amongst themselves. Not singing, says one commentator helpfully, but another believes they are 'literally crying out for help'.[14]

Perhaps I was too hasty in my previous essay to dismiss sentience in the vegetable kingdom. The idea of a 'sensitive' plant may be more than a poetic allegory. If they can communicate with each other, and with animals, then it's not a big stretch to say they have feelings. Things are certainly getting curiouser and curiouser. Or perhaps just dumber and dumber.

First up, plants do make sounds. A few years ago,[15] I interviewed Melbourne scientist turned music composer, the late Martin Friedel, about how he converted the sound of clicking sap, burrowing insects, singing birds, swaying branches and even base notes from water moving through roots into music for human ears. Friedel took his highly sensitive microphones into forests and botanic gardens, capturing every utterance he could associate with a tree.

That made sense to me. You can hear most of these noises if you listen carefully. On the other hand, the plaintive cries and complaints captured in recent experiments are what we call ultrasonic, which means the frequency of the sound is above what most humans can detect. A good microphone will pick them up, but it doesn't make for very interesting listening, given we still can't hear anything.

If we could, we would hear a an occasional (ultrasonic) blip – maybe one per hour – from a healthy, and apparently relaxed, plant. If you drop the pitch and slow these noises down, they sound like popcorn popping or a knuckle cracking. Add some stress to the life of a tomato, tobacco, wheat, maize, cactus or grape vine (just some of the plants tested) and they become decidedly chatty.[16] More like 35 or 40 crackles per hour.

That inaudible pop or crackle is thought to be the sound of an air bubble forming or breaking inside the xylem, the tubes carrying water and nutrients from the roots to the leaves. I remember Martin explaining this to me when I asked him about how something without a mouth or obvious musical architecture could create any of its own noises. Bubbles form under water stress or perhaps when the xylem is severed during pruning (no doubt a 'stressful' situation for a plant), he said. Which makes sense here.

We know plants respond to change (stress) by changing colour, aromas and shape: autumn leaves turn red partly to warn off predators, freshly cut grass aromatics may deter insects, and roots can thicken in response to underground vibrations. But do they deliberately make sounds to illicit a response in animals or even other plants. Bats, mice and moths could certainly hear such ultrasonic sounds, but would they care? Well, maybe.

An evening primrose, *Oenothera drummondii*, can[17] produce sweeter nectar within 3 minutes of being exposed to the sound of a bee. Three minutes! In this case the petals of the flower act as the 'ear', vibrating to the sound of a flying bee or anything of that same frequency. The rattling of the petals is apparently closely correlated with a rapid production of sweeter nectar.

There is also a rainforest vine from Cuba, *Marcgravia evenia*, which has bowl-shaped leaves to reflect sonar calls back to pollinating bats, helping them to locate the flowers. And a pitcher plant from Borneo,

Nepenthes hemsleyana, does much the same to lure bats to its traps, where they defecate and leave much-needed nutrients. These are just a couple of examples[18] of many plants now known to wield some form of 'acoustic reflector'.

It is more difficult to think of a purpose, or value, to what are effectively burps and farts. The sounds are so infrequent and faint that even a creature with the right kind of hearing is unlikely to discern them above the background noises of the night or day. Possibly, just possibly, a moth could use the ultrasonic popping sounds to find the best place to lay its eggs – presumably a healthy plant would provide more food for its offspring. This is a long way from being proven, but it's a reasonable idea to test.

As for other plants 'listening' and responding, which part of a plant would be picking these signals up? It would be odd for flowers to be acting as ears in this case, given they are transient features on most plants. So far there is no proof for any response from neighbouring plants, so let's put that one into the fanciful category for now.

Gardening Australia, February 2025[19]

Touchy subject

Is a caress better than a stake?

Dogs seem to enjoy being patted. Our cocker spaniel, Reg, has an apparently endless longing for tummy rubs and back scratches. Cats, I'm told, may get some pleasure from being stroked. Plants, on the other hand, generally don't like being touched. At least that's the conclusion from recent studies into 'thigmomorphogenesis' – more on that absurd word later – which perversely might also explain why stamping on plants is better than staking them!

It seems that repeatedly caressing your leafy friend may slow its growth by nearly a third.[20] Even gentle petting can disrupt the assembly line inside a plant cell, stopping up to one in 10 of the plant's

active genes from providing the instructions needed to build proteins and other essential products. This in turn stunts the plant.

It doesn't take much. The gentle brush of your hand, maybe one plant swishing against another, or even the flick of a dog's tail while you are patting the other end perhaps. Spraying water onto leaves or the sudden passing of a shadow can trigger the same response. A response with that intimidating name.

Back in the 1970s, the North American plant physiologist, Mordecai (Mark) Jaffe, decided we needed a way to describe the way his thale cress, *Arabidopsis thaliana*, behaved when it was touched. He settled on a word combining *thigma*, to touch, *morpho*, its shape and *genesis*, for the way it changed. In his experiments on thale cress and then other crop plants, thigmomorphogenesis produced shorter, stouter plants, often with fewer leaves or flowers.

Farmers in Japan already had a simpler word for thigmomorphogenesis,[21] and a very practical application for its use. Mugifumi – *mugi* for wheat or barley, and *fumi* for treading – described the stamping on grain seedlings to both strengthen their roots and encourage shorter, thicker and stronger stems. In this case, the result was an increase in yield because the stunted plants were more robust and bore more seed.

It might seem odd for a plant to slow its growth when touched, but there are a few theories on why this reaction may have evolved. One is in response to repeated herbivore attack. If the first round of insect attack triggered stunted and less vigorous growth, that plant would be less appealing to next season's cohort of insects, who would instead seek out healthier-looking plants.

If this theory is true, those individuals with a tendency to thigmomorphogenesis – or mugifumi, if you prefer – should in the long run survive longer and reproduce more successfully than those that without it, despite their reduced overall growth.

Alternatively, the plant may be misreading an ancient cue to 'toughen up' when buffeted by strong wind or grazing animals. We know that wind stress on a young sapling can encourage more roots and a stronger tree trunk.[22] Stress in this case causes the plant to switch on additional genes, including some that code for compounds speeding up the production of lignin. Lignin, of course, is the material that gives the wood in trunks and branches its rigidity.

More wind, and more lignin, leads to a thicker and shorter trunk, presumably because more energy goes into lignin production than cell size or number. You can see a dramatic example of this effect in the shrubs and trees of alpine areas, with their stunted form due to persistent strong winds.

For the home gardener, there is a lesson in all this. Staking a tree may give you a taller tree with more canopy, but it is likely to be less stable in strong winds due to a weaker trunk and poorer root growth. As any garden guru will tell you, generally, it's better not to stake.

Sometimes, though, a compromise is needed. If you want an upright tree in an extremely windy place you may have to do a little staking while young. Limit the stake to two-thirds of the tree's height and attach it loosely so the stem can bend a little in the wind. That way you'll still get the benefit of some extra lignin. It doesn't matter so much with annuals and (annually reshooting) perennials where the stems are not, or less, woody and they will die back after a year anyway.

The way I see it, if the caress of a human hand is really misinterpreted by a plant as a buffeting wind, you can eschew the stake entirely and give the trunk a gentle pat, or a nudge, every now and then.

Gardening Australia, March 2025[23]

Breath of fresh air

Plants indoors and out will make you healthier, happier and smarter

We all know gardening is good for us. Time spent outdoors digging and weeding makes us happy and healthy, and more than likely we'll live longer than we otherwise might.[24] Gardening – indoors or out – also has some very particular and unusual benefits.

There's a microorganism in soil and surrounding air that has been compared to Prozac, in a good way. I heard of it through Lalage Snow's confronting but inspirational book, *War Gardens*, about how gardens help people surviving in conflict zones.

This connection was first noticed when an injection of *Mycobacterium vaccae*, a common bacterium in garden soil, made cancer patients happier and less stressed. It did the same for mice in laboratory tests, giving them the edge over other rodents when navigating mazes. Mice, and humans, can get those same benefits from digesting the bacterium, so imbibing a little garden air is a good thing.

Gardening inside the house may not expose you to as much of this stress-relieving bug but it can make your home a healthier place to live in. Take the air you breathe. At times there can be too much carbon dioxide inside our homes but thankfully this is a gas that plants need, turning it into more welcome oxygen through photosynthesis. (And don't worry about what happens at night when plants switch from absorbing carbon dioxide to producing it – as I explain in the Postscript, the amounts are negligible.)

With this gas exchange in mind, the more light you provide for your indoor plants, the more carbon dioxide they convert into oxygen. Consider installing a skylight or add a few extra LED globes. You might try what we have in our home, an 'LED skylight': a ceiling light panel powered by solar energy.

Healthy plants will also increase the humidity in a room, helping to relieve dry eyes and some skin irritations. They do this through transpiration, where the water you deliver to the pot is drawn up through the plant and released through tiny pores in the leaves. This keeps the plant cool but also provides water for the photosynthesis taking place within the leaves.

Plants can also help with a group of unwanted chemicals in the home called volatile organic compounds (VOCs). These are carbon-based chemicals (hence 'organic') that become airborne (volatile) at room temperature. They are released from furniture, carpets, paints, cleaning fluids and pretty much all electrical equipment.

Plants soak up VOCs when they draw in that carbon dioxide through their leaf pores, as well as absorbing various chemicals into the waxes on their leaves. The plants themselves, and their pots, produce their own VOCs – but, it seems, less than they take up.

And it's not all about the plant. The soil in your pot, with its rich ecosystem of microorganisms, can absorb as much, or more, of these

VOCs. On balance, you'd expect a healthy, live plant in a good potting mix to give you the best result.

The downside is that you need a lot of them to do any good: botanical writer James Wong estimated[25] he would need about 5,000 plants in his small London flat to provide the same improvement in air quality as an open window. Remember too, that his flat is in the middle of one of the world's most polluted cities!

There is also some dispute about which plants are best for removing nasty chemicals indoors, and despite some specific recommendations arising from NASA supported studies, it is hard to pick winners. Ferns and herbs have been suggested, with softer foliage presumed to be better than woody plants. Any plant requiring less light will grow more actively indoors and therefore is most likely take up more chemicals. But be wary of advice any more specific than that.

Regardless of how much indoor plants contribute to purifying our air, there are good studies showing that hospital patients *feel* better and need *less* medication with plants in the room. The only caveat here, and I'm sorry to have to mention this, is that even plastic plants can help with healing. Just the presence of something plant-like makes us feel and actually get better. That's a good thing of course where a room, such as in a hospital, must be kept sterile to protect the most vulnerable from infection.

We would all agree that having plants around (or in extreme cases things that look like plants) is a good thing. If you can combine that with a little physical activity, all the better. I recall being told that potting up a plant is more relaxing than watching television (although maybe you can get a double benefit from potting while watching *ABC TV Gardening Australia*).

But if even that is too much effort, due to poor health or other circumstances, find a place to sit by a window overlooking a garden. Simply looking at living green spaces has been shown to boost concentration, lower blood pressure and help you recover from illness and injury. I can guarantee you will be healthier, happier and smarter.

Gardening Australia, December 2020[26]

Postscript

The year before this story was published in *Gardening Australia*, I included a selection of 'Fact and Fiction' snippets, a short item about why it's OK to share your hospital (or bed-) room with living plants.[27]

As I wrote then, celestial bodies aside, should we share our bedrooms with plants at night? During the day, plants release oxygen as a by-product of using the Sun's energy to produce sugars from carbon dioxide and water (photosynthesis). They also generate a small amount of carbon dioxide when they break down those same sugars to grow and thrive (respiration), but in daylight plants are overwhelmingly net producers of oxygen.

In darkness the balance switches to consuming oxygen as they continue to respire but don't photosynthesise. This has led some to recommend that hospitals clear their rooms of all plant material at night.

Yet, like the effect of the Moon's light or gravity on seedlings, this switch to consuming oxygen is negligible. Plants do take in oxygen and produce carbon dioxide in the dark, but then so do we! I'm sure many of you have slept in a room with another person, an organism that also respires, and you'll notice you didn't die from lack of oxygen. Even if the windows are closed and the room is not air-conditioned.

In fact, a typical plant respires far less than a human. You and I need about 50 litres of oxygen every hour while resting. A small pot plant consumes around one-fiftieth of a litre of oxygen every hour at night. No need to worry about a plant or two in your room at night then. You'll get all those benefits of fresh oxygen in the day, removal of nasty chemicals from the air 24/7 and the proven reduction in stress levels associated with spending time with plants.

And yes, you can take flowers to someone in hospital, outside intensive care and other quarantined areas where you'll have to make do with plastic plants. Either way, it will make your loved ones feel happy and heal quicker.

Nature by numbers

The Fibonacci sequence may not be as common in nature, architecture and garden design as we imagine

What is the connection between those swirling spirals of seed in a sunflower head, the off-kilter symmetry of pinecones and our fondness for odd-numbered plantings? Maybe none as it turns out, but we can track our obsession with these patterns back to at least the 13th-century, to accountant and mathematician, Leonardo of Pisa.

Fibonacci (as he was nicknamed a few centuries later) loved a good conundrum, including the number of rabbits you might expect from a single breeding pair. As we know in Australia, it doesn't take long for a few bunnies to become a plague, but our friend from Pisa was interested in the detail of what happens next.

By Fibonacci's reckoning, the first couple of rabbits will beget two more pairs, followed by 3, 5, 8, 13, 21 and so on. The 'so on' is the sum of the two previous numbers: the series starts with one breeding pair, so the next number is also one (zero + one), then two (one + one), and we are off!

We now call these Fibonacci numbers, and they *seem* to pop up all over the place in nature, architecture and art. As does the golden ratio, also known as the divine proportion or simply phi, which is what you get (roughly) when you divide a Fibonacci number by the one before it. That number, 1.6180339887498948420..., continues forever with no discernible pattern (making it a so-called irrational number).

The quotient of adjacent numbers in the Fibonacci series oscillates above and below this number as you progress along the series, and the larger the pair of Fibonacci numbers the closer the approximation. For example, 55 divided by 34 (1.617647...) is a better representation of the golden ratio than 8 divided by 5 (1.6).

Now, back to nature. The seeds in a sunflower head or the scales on a pinecone are arranged in rows that spiral out from the middle of the flower or the tip of the cone. You should be able to see a clockwise spiral and an anticlockwise one. If you count the number of spirals in each direction it is often a Fibonacci number. The number of spirals in one direction is usually the Fibonacci number before, or after, the number of spirals in the other.

Which is very cute. You can also find Fibonacci numbers, or rough approximations to them, in a rosette succulent such as an agave, aloe or echeveria, or in a head of Romanesco broccoli. If you count the number of spirals, *more often than not* it will be a number from that series.

Again, very cute, but the reason for the pattern is rather prosaic. The most efficient and competitive place for a new leaf is to be offset by 137.5 or 222.5 degrees to the last leaf, rather than directly adjacent or opposite. This results in as many leaves as possible accessing full sunlight. Over time this offset creates spirals of leaves, and in the most efficient packing there will often be a Fibonacci number of them.

But not always. While an overwhelming majority of sunflower heads have 34, 55 or 89 rows of seed spiralling in one direction – all of which are numbers in the Fibonacci series (*0*, *1*, 1, 2, 3, 5, 8, 13, 21, 34, 55, 89...) – nearly one in five don't. Not to fear though: some of these non-Fibonacci compliant flowerheads have seeds spiralling in numbers that match other mathematical series, based on two different starting numbers.[28]

It could be a number from the Lucas series (*2*, *1*, 3, 4, 7, 11, 18, 29, 47, 76, 123...), named in honour of Edouard Lucus, the man who coined the term Fibonacci numbers. Or the F4 series (*3*, *1*, 4, 5, 9, 14, 23, 37, 60, 97...), named, I assume, after the tally of the first two numbers. Even from the more obviously named Double Fibonacci series (*0*, *2*, 2, 4, 6, 10, 16, 26, 42, 68 ...).

Seed arrangements that happen to match one of the numbers in these series, or even others such as a Fibonacci number plus or minus 1 (or sometimes more), are granted special significance by some. Given there are few numbers left outside the combined data sets, I find it hard to get too excited, although I'm told the early development of floral spirals in each flowerhead does strictly adhere to the Fibonacci pattern, with the later amalgamation of some parts leading to alternative arrangements.[29]

More dubiously, it is sometimes stated that the number of petals in a flower or flower-head is also a Fibonacci number, again citing a sunflower as an exemplar. Yet in a study of 1,000 sunflower flower-heads, it seems just over a third had 21 petals, a few 13 and the rest anywhere between 12 and 31.[30]

In architecture, the design of the Acropolis in Athens and the Parthenon in Rome are sometimes said to indebted to the golden ratio.

In the case of the Acropolis, the dimensions don't quite match up and I gather there is no evidence the original design of that or other buildings of antiquity took this proportion into account. On the other hand, Renaissance buildings such as the Laurentian Library in Florence and the Notre-Dame de Paris do have design elements based on the golden ratio, as have some Renaissance gardens.

In design more generally, an odd number of something is considered more attractive than an even number if we want a relaxed and informal feel. In small gardens this means we tend to use Fibonacci numbers like three and five when planting groups of trees. But lest you get too excited about this correlation, two and eight are also Fibonacci numbers.

A mix of shrubs that reach 8 metres mixed with some that reach 5 – both Fibonacci numbers – is a planning tool used by some landscapers, but again it's hard to give this any deeper connection to the whole Fibonacci series.

You'll find the golden ratio referenced in the shape of garden beds. A rectangular bed with proportions approaching 1 to 1.6, say 5 metres wide and 8 metres long, will look 'about right' to most people. Of course, if you have a long narrow garden or a curvy space you'll have to either break it up into little golden-ratio units, or create something with a different aesthetic.

The Fibonacci number and the golden ratio may well provide for an intrinsically beautiful object or collection of objects, but in most cases other practicalities and other equally attractive proportions prevail. In nature, Leonardo of Pisa's calculations didn't actually work in the rabbit burrow – he failed to allow for more than two offspring for a start – and they are only a rough approximation in flower and plant geometry. Still, the mathematics remains beautiful.

Gardening Australia, January 2019[31]

Postscript

I wrote this essay a year or so after I got extensive feedback from an interview and website story for *ABC RN Blueprint for Living* in 2017.[32]

I was able to correct a couple of minor historical facts and to be even more circumspect and sceptical than I was already about golden ratios being found in nature, often or ever.

It remains a contentious subject, at least in my mind. A recent media release and scientific publication[33] from the University of Edinburgh reported on a 400-million-year-old clubmoss fossil with a sequence of leaf development that – if I read it correctly – is 1, 2, 3, 4, 5 and so on, so not a Fibonacci sequence.

This doesn't surprise me, but I note that the media story says that Fibonacci spirals make up over 90 per cent of the spirals found in nature – citing 'sunflower heads, pinecones, pineapples and succulent houseplants'. All of these, I understand, are *more or less* Fibonacci but not always strictly so. Still, they are presumably *more so* than that ancient clubmoss.

It's flowering time

Plants might flower like clockwork, but you can't set your watch by them

Melbourne is proud of its floral clock, part of a 19th-century sensation reaching Australia in the 1960s. Thanks to a buried, concrete-encased, synchronous motor – Swiss made – the giant hour, minute and second hands glide over a circular garden of begonia, marigold and other annuals. In some years, long-lived box hedges have been added to provide better definition. Beautiful – at least to some eyes – and reasonably functional.

The original concept of a floral clock was quite different, and quite barmy. By the mid-18th century, natural philosophers (the word 'scientist' was not yet coined) were starting to get a grip on the variety and apparent vagaries of plant reproduction. Different plant species not only have different-looking flowers but those flowers bloom at different times of the year.

Some plants even open and shut their flowers at particular times during the day (or night) to take advantage of, say, morning or evening pollinators. This led Swedish biologist Carl Linnaeus,[34] creator of what we call our binomial system of naming (e.g. *Homo sapiens*, for us humans), to postulate a *horologium florae*.

With judicious planting, one can *imagine* a garden where the open flowers tell you the time day. In the Swedish city of Uppsala, where this idea first took root, you might bounce out of bed when the dandelion (*Taraxacum officinale*) flowerhead unfurls at 6 a.m., rush to work when it shuts between 8 and 10 a.m., grab a sandwich at noon when the field marigold (*Calendula arvensis*) flowerheads close, leave work at 6 p.m. when the sad geranium (*Pelargonium triste*) finally opens its dull yellow flowers and, if you like, party until the Queen of the Night cactus (*Selenicereus grandiflorus*) blooms at midnight.[35]

Let's get something clear up front, though. This is like picking your dream sports team of all time or forming the world's best supergroup.

It is a paper or mental exercise. A true floral clock can't be created in the real world. These plants won't grow together in the same place, they won't all flower on the same day of the year (each flower of the Queen of the Night cactus, for example, opens for one night only), you would have to recalibrate for every degree of latitude or altitude, and depending on the weather some might not open at all.

While a full functioning floral clock is a folly, you could conceivably create a coarse local variant that might work in spring or thereabouts. You could start with two Victorian species of flax lily (*Dianella*), one of which (*Dianella amoena*) opens early to mid-morning, the other (*Dianella tarda*) early to mid-afternoon. But then you know when it's mid-morning or mid-afternoon, don't you? Better to spend your day pondering how plants keep time.

Most of them set their clocks from the amount and quality of the daylight they receive. They have a system for tracking seasonal changes in day length, triggering when they flower or renew growth. The variations within a single day are controlled by light, temperature and even humidity, plus an in-built circadian rhythm like animals. Such precision is the result of evolution, with successful plants protecting their assets (flowers) until a suitable pollinator is active. For the night-flowering cactus this is a moth. For the two flax lilies, it may be the same or different species of native bees.

Some plants will open and close their flowers on a daily cycle, by growing new cells every day in the inside (to open) or outside (to close), or by expanding and contracting existing cells in the petals. It is worth the effort to protect delicate parts from wind, rain and dew, and to keep pollen dry and ready for action.

Plants growing in tough environments are more likely to have these diurnal cycles. A good example is pigface (*Carpobrotus glaucescens*), a succulent adapted to dry sand dunes. It has daisy-like flowers that close every night. The sensitive plant (*Mimosa pudica*) folds its leaves at night to protect that equally important, and in this case sensitive, part of the plant.

The flowering head of a sunflower (*Helianthus annuus*) takes it all a step further, arcing during the day in response to its own internal clock. Young sunflower plants (with their flowerheads unopened) track the sun from it rising in the east to its setting in the west, then get themselves sorted overnight – turning their heads back to the east – so they can do the same the next day. Once the giant yellow flowerhead opens, it stops moving about and faces east.

The sunflower has a circadian mechanism behind its solar tracking and then some internal signalling to position its mature flowerheads eastward. The solar tracking is caused by different growth rates on the sunny and shady side of the stem. During the day the cells on the east side grow a little faster than those on the west, and then vice versa at night. This comes to a halt when the stem is fully elongated, leaving the open flower facing the morning sun, where warm flowers attract more bees.[36]

Plants keep time in other ways. Flowers can last a few hours (our night-flowering cactus) or a few months (moth orchids), and the entire lifecycle can be a matter of a few months (desert annuals) to thousands of years (the bristlecone pine). Annuals typically flower once and die, all within 12 months, but other plants, such as the century plant (*Agave americana*) and several palms, will live for a few decades then flower once and die. For most woody plants and trees though, they flower each year and continue growing until like us they expire.

Nearly all plants have seasonal cycles, of course, although these can be manipulated to varying extents. For example, in mid-winter, poinsettia produces red or pink leaf-like bracts around its tiny flowers, making it the perfect Christmas bloom in Europe or North America. They need long 'dark periods' of about 3 months for them to flower and for the bracts to become intensely coloured.

Northern Hemisphere gardeners report that interrupting the 'long night' by even flashing with a torch during October through to December may stop the leaves changing colour. To meet (or perhaps create) demand for flowering poinsettias for the Southern Hemisphere

Christmas, in summer nurseries create artificially long nights inside a glasshouse.

While this sounds highly controlled, a colleague at Royal Botanic Gardens Victoria had a pot plant growing an east-facing windowsill that flowered happily in mid-summer. Maybe that was caused by the artificial lights or shadowing on the sill. In any case, it seems you can't really set your watch, or your Christmas shopping schedule, from a flower.

Talking Plants, January 2018[37]

Postscript

I posted a follow-up piece to this essay 4 years later, after photographing a paper daisy at Royal Botanic Gardens Cranbourne.[38] The plant was a golden everlasting (*Xerochrysum bracteatum*), with flowerheads closed or just a little ajar at 7.30 a.m., when I took my first walk around the Australian Garden, then fully open at 10 a.m. It was a sunny day so I couldn't be sure if the opening was triggered by warmth, the amount of sunlight reaching the flower or some more complicated circadian cycle.

It got me thinking a little more about *why* plants might have these daily flowering cycles. You would expect it has something to do with pollinators, and mostly it does. A 2016 study reported in the journal *Science*[39] cleared things up for the sunflower. They tested flowering plants in pots, in growth chambers and in paddocks, restraining flowering stalks with stakes to stop them reorientating.

While the orientation of the sunflower head does respond to sunlight, there is an internal clock that can override a lack of external stimulants for at least a few days. This system of genes and signalling chemicals sets up the flowers to warm up as quickly as they can in the morning, making them more attractive to passing bees.

The 11 o'clock flower (or rose moss, *Portulaca grandiflora*) also relies on bees, so its flowers are triggered by sunlight to open fully in late morning, when the pollinating bees are at their most active. On a cloudy day, the flowers remain closed, as are those of many sun orchids (*Thelymitra* species) in Australia. Why waste precious resources opening flowers when your pollinators are scarce?

For the aptly named 4 o'clock (p.m.) plant (or marvel of Peru, *Mirabilis jalapa*), that means waiting until later in the day to unfurl your floral hoarding. That's when the hawk moths become active. While there may be insects around earlier in the day, those are less effective pollinators of these flowers. Similarly, night-flowering plants produce the goods when bats, moths and other nocturnal creatures are out and about.

Yet another example of how the structure and functioning of a flower is shaped by its coevolved pollinators.

Hungry leaves

There is more than one way to feed a plant

I was surprised to learn that plants take up water and nutrients through their leaves. Plants have a very orderly plumbing system. There is one set of pipes to take water and nutrients from the roots to the leaves and another to carry supplies from the leaves to the rest of the plant.

These pipes are what we call the vascular system, and in the leaves, veins. Their cargo is primarily sugars, produced when sunlight hitting the leaves powers the conversion of water and carbon dioxide into this essential plant food. The liquid travelling up from the roots is a solution of various minerals, such as nitrogen, phosphorus and magnesium – essential building blocks for the sugar factories, the leaves.

All very simple: nutrients go up, sugars go down. Leading to the obvious conclusion that to fertilise a plant you should add your potions and pellets to the base of the plant. That's why there are all those products on the market for improving soil fertility.

So what's the story with foliar sprays? Isn't that working against nature? Well, we've known for a long time that there are exceptions to the up–down system. In 1679, around the time Sir Isaac Newton was unravelling the physical properties of light and gravity, Edme Mariotte, a French physicist, announced that plants can absorb water through their leaves.

It took another two centuries to demonstrate that minerals can also enter the plant this way, and until recent decades to clarify the mechanism. It is now known that water and minerals can pass through the cuticle, the waxy outer protective layer of a leaf, as well as through the tiny pores in the leaf surface called stomata. In nature this presumably provides a way for the plant to absorb additional nutrients from rainwater.

As a gardener you might say who cares about the science, just tell me the best place to apply my fertiliser – leaves or soil? The answer is both, depending on what ails your plant, how quickly you want the mineral absorbed and how much time you have to apply the treatment.

If your plant needs one of the more 'mobile' nutrients, such as nitrogen or phosphorus, either method should be fine. If, on the other hand you need to add something like calcium, which travels well in the water pipe but not so well with the sugars, you'll need to direct your application very carefully or revert to feeding the roots.

Minerals applied to the root zone will make their way up the plant, and a sprinkling of granules or heavy watering with a rich nutrient soup may be quicker to administer, but you do risk it all washing or draining away. Not only is that potentially wasteful and expensive but it can lead to detrimental effects downstream.

The research clearly supports foliar spraying as the most efficient way to get plant minerals to a particular part of a plant. Use your favourite liquid fertiliser but dilute the dosage to about a quarter of what they recommend for adding to soil – high concentrations of minerals won't be absorbed and may burn the foliage.

For mobile nutrients it doesn't matter where you spray the plant: nutrients will be absorbed through buds, flowers, fruits and even the trunk. But don't, as I've seen it described, 'spray and pray'. One of the benefits of applying your fertiliser direct is to target the part of the plant you wish to encourage or cure. For example, adding boron to banana fruits directly can increase their weight and size.

For greatest impact, spray when stomata are likely to be open. For most plants that would be during daylight hours but not the hottest part of the day, when stomata may close to reduce water loss. Ideally chose a warm, moist and calm day, with rain not predicted for at least 24 hours.

A few drops of a surfactant such as detergent will help the water adhere and spread across the leaf. You can reapply as often as necessary to cure the symptoms or increase your yield, but once a week is a good rule of thumb.

Don't worry about fancy equipment. Use any bottle with an atomising spray and apply until you see the water run off the leaf. After that you are applying a soil fertiliser!

Gardening Australia, February 2018[40]

Postscript

Nothing much has changed since I wrote this in 2018. Gardeners are still reluctant to 'water leaves' but the experts say it is one of the best ways to fertilise a plant. According to a recent review,[41] the application of fertilisers directly to leaves rather than through the soil is increasing in agriculture. That is because you can achieve the same results quicker and with less fertiliser, while causing less damage to the soil – excess fertiliser will change soil acidity, salinity and chemical composition.

There is that caveat though. Not all plant nutrients are best applied through the leaves. Some, like selenium, may be lost through transpiration and others, such as calcium (mentioned above) and boron, are not transported through the plant's phloem. Generally, though, fertilising through plant leaves should be good for the plant and the environment.

The burning question

Watering plants in the middle of the day is unlikely to damage leaves or start a bushfire

I was a country kid. When I wasn't trapping sparrows under propped-up tin cans or building Lego prisons for grasshoppers, I was irritating ants with a sunbeam focused through my magnifying glass. That magnifying glass was also good for scorching, and sometimes burning, paper and dry leaves.

All fires were quenched and most animals escaped unharmed, but I did learn some basic physics during these escapades. From my experiments, I learnt that a convex-shaped, transparent object could focus the sun's radiant heat into a tiny dot, and that tiny dot could start a fire.

Stories abound of water droplets doing just that, should they remain too long on a leaf or on our skin. The fear is mostly of

discoloration on leaf or skin, but some have claimed that a bushfire might start with a freakish drop of water on a dry plant.

All very plausible to my childhood self. Indeed, the 'scorching water droplet' hypothesis, if I can call it that, is repeated on about three-quarters of all horticultural websites. But is it true?

Thanks to clever research in Hungary, followed up by neat school experiments in the same country, I can confidently say that most leaves (and skin) are extremely unlikely to be burnt by water droplets.[42]

The tests showed that glass beads, simulating water drops, have to remain on the leaf surface for more than half an hour before there is even a chance of focused sunlight burning the leaf. On a day with enough sunlight to burn leaves, real water droplets will usually evaporate in less than 25 minutes. While in contact with the leaf, they keep it cool as well.

In any case, water droplets on most leaves are quite flattened. I remember from my childhood that the shape of the lens and its distance from the ant was very important. Only when sunlight was focused on a small point did the ant heat up, causing it to jump away. For a plump, roundish lens, that point was close by, and for a lens shaped like two dinner plates glued together, the burning point was much further away. Our flattened water droplets are usually too close to the leaf to cause any damage.

Late or early in the day, when the light shines through the side of the droplet, it could conceivably focus light closer to the surface, but at those times the intensity of the sunlight would be too weak to cause damage.

More spherical water droplets might form on the water-repelling surfaces of ginkgos and water lotus, but they generally roll off. Only by raising the droplet above the leaf surface do you shape a better lens, focusing the light directly onto the plant tissue and keeping the cooling water away from the target site.

Hairy plants can do this, particularly if the hairs are waxy and repel water. It has been seen in the floating fern *Salvinia molesta*, but I expect things like lamb's ears (*Stachys byzantina*) or the native woolly cloak-fern (*Cheilanthes lasiophylla*) might suffer the same fate. On such long-haired plants, a water droplet could conceivably sit aloft for long enough and far enough above the leaf to cause burning.

For hairless or mildly hairy plants, watering in the middle of the day won't burn their leaves. Brown spots are more likely to be due to impurities in the water such as chlorine, salt or a high concentration of fertiliser. There could also be some physiological stress associated with spraying very cold water onto a very hot plant surface, but that's even less likely.

For a hirsute plant, I'd recommend you avoid watering onto foliage in full sunlight. Whether the leaves are damaged or not will depend on where the droplets end up in relation to the leaf surface and the angle of the sun, and how quickly they evaporate.

Still, if your plant is wilting and you can't avoid spraying the leaves, I'd risk it. As to fear of rain igniting dry vegetation, I think you should get over that one. Even for hairy plants, the moisture would most likely dampen the fuel load for as long as it took strategically suspended droplets to evaporate. Like sunburn on plants, there are far more likely causes of bushfire.

Gardening Australia, November 2018[43]

Postscript

This essay was written in response to questions about the previous essay on fertilising through leaves. Should we still avoid watering plants in the middle of the day, I was asked? The answer, as you can see, is 'well yes, but …'.

I have watered plants in the middle of the day, when it was particularly hot and I had forgotten to water the evening before or that morning, or when one of my 'indicator plants' (such as hydrangea) was looking very sad, with leaves and tips drooping. It isn't an ideal time to water, but sometimes needs must.

The key message here is that as far as possible, water your garden in the cooler parts of the day – most experts would suggest early morning, so you don't leave fungal-attracting moisture on the plants overnight – but not to hold back if a plant looks like it needs a refresh in the middle of the day. Yes, the watering will be less efficient but it's highly unlikely you'll burn their leaves.

True blue

Alchemy in the garden has its limits when you want a blue hydrangea

Most gardeners know that increasing the acidity of soil can change the colour of some hydrangea flowers from pink to blue. It doesn't always

work, because there is more to it than acidity and, to borrow from the old Castrol slogan, 'hydrangeas ain't hydrangeas'.

Acidity is important in the expression of flower colour of many plants. While there are lots of pigments involved, the pinks and blues are mostly due to anthocyanins. These anthocyanins slosh around inside vacuoles, watery storage compartments inside plant cells, and their colour depends on the acidity of the solution inside that vacuole.

When petunias are genetically modified in a laboratory to produce a less acidic vacuole solution, the flowers change from pink to blue–purple, a colour not found in nature. This is despite the amount of anthocyanin in the vacuole remaining the same.

In hydrangeas, though, things are different. It is the amount of aluminium, rather than acidity, inside the vacuole that changes the colour of the anthocyanin. For most larger-leaved species (e.g. *Hydrangea macrophylla*), the flowers are blue when aluminium is readily available and pink when not. I should note here that hydrangea flowers don't have true petals, so it's the sepals (the typically green outer layer of floral parts) that appear colourful and petal-like.

The intensity of that flower/sepal colour depends on both the amount of aluminium available in the soil and the amount of anthocyanin in the flower (if there is none, the flower is white). When there is plenty of aluminium in the soil, the flowers will almost always be a shade of blue.

In soil with less aluminium, the acidity becomes important. If you want blue flowers, add aluminium sulfate, sometimes called alum, to the soil to make it more acidic. The standard measure of acidity, pH, should be less than 6.5.

Coffee grounds work as well, which is why you need to be careful spreading coffee on your garden bed if you have plants that don't like acidic soil. Similarly with compost and certain mulches, on both counts.

Adding lime to soil makes it more alkaline, binding up the aluminium so there is less available for plants. All other things being equal, this means you will get pink hydrangea flowers. Be careful though. If the soil is packed full of aluminium or is naturally acidic, you may kill the plant (e.g. from a shortage of available iron) before it produces pink flowers!

Unhelpfully, the familiar litmus paper used to measure pH turns blue when a solution is *alkaline* and red when *acidic*, the opposite to hydrangea flowers. Equally unhelpfully, not all hydrangeas respond to fiddling with the soil chemistry. For example, the oak-leaved hydrangea – *Hydrangea quercifolia* – always has white flowers with a tinge of pink. Maybe the vacuole in the sepals is too well buffered, or perhaps the pigments are of a kind that is not influenced by acidity. In any case, you won't be able to change their colour with a handful of lime or alum.

I also grow the so-called blue lacecap hydrangea, a species of *Hydrangea* that used to be classified in a separate genus called *Dichroa*. I assumed it would hold true to its common name no matter what the soil chemistry. Yet when those pretty, white buds unfurl in January, they reveal what I can only describe as pinkish-purple flowers. Definitely not blue.

This is contrary to what you'll read in otherwise reliable sources such as the websites of Missouri Botanical Garden or Flora of China. Although the old genus name – *Dichroa* – should have been a clue ('Dichroa' means 'twice colour'). Horticultural icon, Jane Edmanson, to her credit, described the colour alternatives when she spoke about this species on ABC TV *Gardening Australia* a few years back.

Since then, I've noticed that a few nurseries are on to it, using the more ambivalent common name – blue/pink cap. It seems that like *Hydrangea macrophylla*, the flower colour of lacecap depends on the amount of aluminium available in the soil. We have only a little in our garden, it seems, and definitely alkaline soil.

Let me finish with some further obfuscation, this time from Trebah Gardens, in a beautiful extremity of Cornwall, United

Kingdom, half an hour or so by hedgerow-lined roads from Falmouth (i.e. a couple of miles). Trebah is a locality mentioned in passing in the Domesday Book of 1086, meaning 'house by the bay'. The Fox family home on this property, built along with the garden in 1838, has views over the garden to a gorgeous bay.

The setting – a deep protected valley down to a sheltered beach – and the collection of 100-year-old rhododendrons, camellias and magnolias, plus a smattering of plants from warmer climes, make Trebah Garden quite different to other grand gardens in England.

Along with some impressive gunneras, the signature plants for many visitors though are the hydrangeas filling the lower valley. I am not aware of *Hydrangea macrophylla* having any invasive tendencies but it certainly flourishes beside the stream here in Trebah Gardens, at least while the gunnera is kept under control upstream (as it must be in the UK, where there are strict conditions on those two retain this aggressive weed in older gardens).

In any case, the thing I noticed on my visit to this garden in July 2012 was the seemingly random mix of pink- and blue-flowered hydrangeas. When I asked our guide how that could be they posited two possibilities: either the pink flowered individuals were grown in specially prepared compost (and they will convert to blue as they settle into their garden position) or there are indeed genetic strains that remain pink or blue no matter what the soil chemistry. I didn't argue but assumed it must be the former.

Returning to London, I began to doubt myself, thinking that if other species of *Hydrangea* could 'hold their colour', why couldn't there be a variant of *Hydrangea macrophylla* that did the same? Sure enough, after a little web surfing I found this on the website for (the now) King Charles III's Highgrove House and Garden: '*Hydrangea macrophylla* "Pia" ... has a dwarf habit growing to 70 cm in height and spread, keeping a good bright pink flower colour no matter what the soil type.'

The hydrangeas at Trebah were definitely not dwarf, but this does support the alternative explanation, where some varieties flower pink no matter what. I'm keeping an open mind on the whole matter.

Talking Plants, June 2009[44]

Postscript

I am not the only one interested in why the flowers of only some *Hydrangea* species and cultivars change colour with soil acidity (and aluminium availability). This is of great interest to nurseries and landscape designers, who would prefer to know with some precision what colour flowers a plant will produce.

A research team in Japan published one of the more recent attempts to unravel the mechanism behind the stability, or not, of flower colour in *Hydrangea*.[45] They confirm that the cultivation of many species in acidic soil, 'containing more soluble aluminium ions', will result in bluer sepals (flowers). As I mentioned above, the bluer colour is correlated with an increase in aluminium in the sepal vacuoles.

They also confirm that 'the effect of soil conditions on the sepal colour differs among cultivars', and that commercial growers would love to find a more efficient and effective way to control this. Noting that some cultivars already have stable red or blue flowers, no matter the soil conditions.

It seems there is more at work here than simply aluminium, and the researchers found a compound called neochlorogenic (3-*O*-caffeoylquinic) acid to also be lower in stable, blue-flowered cultivars.

Apparently, neochlorogenic acid inhibits another form of the same compound, chlorogenic (5-*O*-caffeoylquinic) acid, and it is that form the plant needs – in conjunction with aluminium and the colour-forming compound anthocyanin – to produce blue flowers. Less neochlorogenic acid means more active chlorogenic acid, which in turn means bluer flowers.

The current thinking, then, is that if breeders select cultivars that can accumulate (or hold) higher amounts of aluminium and chlorogenic acid, but less neochlorogenic acid, they will have more chance of producing *true blue* cultivars.

Until then, it's alum or lime, and a bit of luck.

Boom and bust

Not only annuals die after their first flowering

Short-lived annuals, like the tiny cress popping up in my garden apparently overnight, are living life in the fast lane. They flower within a few days, and then ... well, I try to weed them out before they set seed and die.

They are what ecologists call 'r-selected species': 'r' for reproduction, because they do a lot of it. They grow quickly, produce copious seeds and sometimes die within weeks.

At the opposite end of the spectrum are plants that grow slowly, producing infrequent seeds but with individuals perhaps living hundreds of years or more. They are termed k-selected species – 'k', slightly obscurely standing for carrying capacity, because this strategy

works best when offspring enter a population at or near what the world around it can sustain.

Of course, there are many plants with strategies somewhere between these two extremes. I want to explore here what we might call the 'lazy cress' lifestyle.

Annuals, by definition, flower and fruit within a year. In the case of the cress, in a matter of weeks or months. If that flowering cycle takes 2 years to complete and then the plant dies, we call it a biennial. And so on. This boom-and-bust approach, with the plant dying soon after it fruits, is typical of a 'r-selected species'.

It is also a strategy employed by a few much longer-lived plants. Most famously, the leafy rosette of the (misnamed) century plant, *Agave americana*, flowers, then dies, after about 20 years or so. The fishtail palm (*Caryota urens*) also grows for about 20 years, then produces flowering stalks from the top down over 10 years before it dies. The *Corypha utan* palm on Cape York takes 30 years. These species are known in the botanical trade as monocarpic – with a single fruiting.

In 2008, a palm was discovered in Madagascar and named *Tahina spectabilis*. At 18 metres tall, with leaves up to 5 metres in width, it is no doubt spectacular. Even more so in flower, I gather, when its massive, one-off flowering event attracts large swarms of insects and birds. Nicknamed the suicide palm, we still don't yet know how many years it takes to flower. And therefore how many years it lives, because we do know if it is monocarpic.

There is a monocarpic legume called the suicide tree (*Tachigali versicolor*). I first learnt about it in Richard Powers' novel, *The Overstory*, but it is a real tree, and it really does die after its first flowering, which is unusual for a woody plant. Unlike the succulents and palms I've mentioned, there doesn't seem to be a 'pup' emerging from the base of the tree to supplement the abundance of seed.

Sometimes a distinction is made for plants where a single stem, not the entire plant, dies after flowering. These are known as hapaxanthic. A good example is the sago palm, which has multiple stems, only some of which set fruit and die in a flowering event.

A character in *The Overstory* sees the suicide tree more as a self-sacrificing tee, creating an opening in the canopy for light and a rotting trunk for nutrients and a convenient nursery for the next generation. An extreme example of self-sacrifice – or a macabre suicide pact – is the mass flowering and death of bamboo. You can see this in your garden or in the native bamboo forests of northern Australia.

While most of the world's 1,200 species of bamboo grow in China and the Americas, there are two, or three, considered native to tropical Australia. The dubious third of these species, *Bambusa arnhemica*, may have arrived here with the First Australians although many consider it a more recent weed.[46] Either way, in an 8-year floral orgy beginning in 1996, over 80 per cent of the Arnhem Land population of this species flowered and died.[47] The dieback was so severe and unexpected, many locals assumed it was part of a weed eradication programme.

Synchronised flowering of bamboos is not uncommon but is usually attributed to plants from the same source and of the same age: propagated vegetatively rather than by seed. In fact it occurs regularly in natural populations, after one to 120 years, but only in some species. This may be an evolutionary adaptation to overcome predators eating all the seed – you save up all your energy and blast out so much seed they can't eat it all – or perhaps triggered by longer-term environmental conditions.

Although considered synchronous, the 8-year span of bamboo flowering in Arnhem Land is consistent with an older origin for this species in Australia. A more recent introduction would result in more genetically similar plants and, as with nursery stock, simultaneous flowering over a shorter period.

In *Bambusa arnhemica*, the bamboo died after flowering, but some bamboo species are only weakened for a while and will soon recover. So they are not truly monocarpic. Other species flower more frequently, some annually, and these will usually continue to grow happily after flowering.

All this reinforces the reason flowering plants have flowers, to literally seed the next generation. There are many ways to do this, but it seems if you do it well enough the first time, it can be best for the next generation if the boomers – sorry, bloomers – move on.

April 2024; Gardening Australia, March 2020; Nature Australia 2004[48]

2

Plants from elsewhere

'Without hard work, nothing grows but weeds'[1]

The wisdom proffered in this tired old saw is predicated on at least two dubious propositions: that weeds are not desirable and that hard work is. I'll leave the work ethic question to spiritual guides and philosophers but I'm happy to tackle the weed matter.

In Australia, weeds are demonised by farmers, environmentalists and a certain kind of gardener. There are good reasons for this stance – as I'll get to later in the chapter – but my starting point is that a weed has no intrinsic ethical merit or demerit.

In *The Bush: Travels in the Heart of Australia*,[2] Australian author Don Watson considers our tolerance of native Australian, exotic and local indigenous plants in gardens, and more broadly, in extra-garden settings ('the bush'). He laments the damage and displacement of native species by rampaging weeds but also the futility of returning land to some imagined original state or trying to create a garden with no impact on the broader environment. And there's the rub.

The most compelling reason for floral exclusivity is to avoid adding to Australia's 'weed problem'. That problem is a big one, with scientists at the CSIRO estimating a cost to Australia of over $4 billion a year in control and lost production.[3] Humans have transported some 28,000 plants species to Australia, about the same number as the natives species that grew here before European and First Australian arrivals. Most of the introductions were deliberate, and more than 2,500 now grow and spread in Australia without our further assistance.[4]

Gardening is – at its most fundamental – the introduction and encouragement of plants we like and the discouragement of those we don't. Encouragement amounts to watering, fertilising, formative pruning and the like. Discouragement usually involves removing offending plants to the point of annihilation of that species within areas under our control. We parse animals in the same way, favouring pollinators and destroying plant predators.

With plants, it's all about selection: choosing those we want to be part of our collection or landscape and rejecting the rest – particularly weeds or poorly performing specimens. While there may be ambiguity around what to do with an attractive, adventitious plant, perhaps arriving by seed, or a weedy plant that just happens to create a useful ground cover, it's generally easy to categorise plants as good and evil within the context of a garden.

On the other side of the fence, and more pertinently beyond the city limits, the morality of plant selection can be vexed. After reading *The New Wild: Why Invasive Species will be Nature's Salvation* by Fred Pearce, I am primed to accept all plants as worthy participants in 'nature'. Yet while I understand Pearce's philosophical and practical conclusion, I still get protective about a rare orchid growing in the clapped-out bush near Castlemaine, or a remarkable seaweed attached to a rocky platform just off the coast. I'm not quite post-modern enough to accept every plant as equal in value. That said, and as you'll read in this chapter, I'm a lot more liberal minded than many of my colleagues and friends.

Let me finish with something you won't find in most gardens, except as an extract or in dried form – a seaweed. Its very name suggests this might be sometime unwanted and detrimental to the environment. Yet these marine algae – as we know them slightly more formally – are an essential part of coastal ecosystems.

I happen to have a one named after me, *Entwisleia bella*, an extremely rare seaweed, confined to an area the size of a standard living room off the coast of Hobart. I'll return to it later in the book but like most gum trees, many terrestrial orchids and half a dozen kangaroo species, this alga is a unique part of Australia's native biota.

It is also seasonal and even at its peak there are no more than a dozen of the red feathery algae attached by a small adhesive clamp to subtidal rocks.

It is rarer and, as it turns out, evolutionarily more distinct (i.e. it has fewer living close relatives) than the Wollemi pine.[5] Personally, I would be very unhappy to hear that the last individuals of *Entwisleia bella* on Earth were eaten by an invasive shellfish or shaded by an aggressive algal weed from elsewhere. I also understand that very few people will ever see this seaweed *in situ* and may not appreciate its beauty or significance the way I do.

That doesn't mean we should knowingly destroy *Entwisleia*, although I accept its conservation may have to be prioritised against other imperatives.[6] I accept also that an 'anything goes' attitude to weeds and pests might lead to the extinction of a species such as this. In a similar way, farming seaweeds for fuel or other products needs to be approached with great caution so we don't repeat the devastation caused by much agriculture on land.

You may not share my concern about this seaweed, but when the Wollemi pine, the crimson spider-orchid, or Gilbert's potaroo are under threat from invasive weeds, our sympathies are seldom with the interloper. There are good scientific, philosophical, empathetic, maybe even spiritual reasons why we should care. The question is, how much?

Drifting on ancient currents

Australia's 'unique' flora is a product of both historical connection and isolation

Our planet is home to about 300,000 plant species.[7] There may be more plant species elsewhere in the Universe but so far we haven't found them. We do know there are more on Earth – possibly tens of thousands – yet to be documented and named by scientists. Those additional species are hiding in jungles and other places remote from where most scientists live, or sometimes even in their collections of pickled and dried plants.

A little over 6 per cent of the named species (19,000 or so) occur 'naturally' in Australia. That is, they are assumed to have been growing here before European settlement in 1788.[8] Six per cent is not a particularly extraordinary proportion for a continent occupying just over 5 per cent of the world's land.

Of course, that plant diversity is unevenly spread, with some smaller areas such as the south-west of the continent being extremely rich, whereas vast areas of the interior are comparatively very species poor. More notably, of those 19,000 (or so) plant species,[9] 92 per cent are found here and nowhere else; they are *endemic* to Australia.

This is why the flora of Australia is considered in a world context to be distinctive or 'unique', despite not being particularly diverse if measured by species per unit area of land. By comparison, Europe has around 3 per cent of the world's plant species growing on just under 7 per cent of the world's land, with 38 per cent endemism.[10] Not much diversity and not much endemism, leading botanists to cheekily call Europe a post-glacial wasteland.[11]

Africa, on the other hand, which is a continent with wide climatic range and conducive evolutionary history, just like Australia, has some 14 per cent of the world's plant species on 20 per cent of the world's land, with 70 per cent endemism.[12] A little under us in terms of diversity, but that can be attributed to its vast size, nearly four times larger than Australia, and, like Australia, having an uneven spread of plant variety. Endemism is again high.

The most species-rich continent, by area, is South America, with 24 per cent of the world's plant species (if you include Central America up to Mexico) on 12 per cent of the world's land.[13] Endemism runs at 60–90 per cent depending on where you are in South America, so not unlike Australia and Africa. These three continents, and other remains of the ancient supercontinent Gondwana, support a large chunk of the plant diversity we see on Earth today.

Such worldly comparisons are imprecise and deceptive, but Australia as a whole has a proportionate number of the world's plant species by area with a disproportionately high level of endemism. That endemism, and the prominence of species such as gum trees and Australian wattles not found elsewhere, gives our flora its special feel. And this may be why we fear – perhaps disproportionately, compared to other nations – the arrival and establishment of plant species from elsewhere.

Yet at least some of our present-day endemics arrived here from overseas, adapting and becoming separate species over millions of years. This includes some well-known plants in a range of plant families, families being the taxonomic category above genera and a handy category[14] for tracking the major evolutionary lineages.[15]

There are about 620 plant families in the world, of which just under half (300) are represented in Australia. So, although only 6 per cent of plant species occur in Australia, close to 50 per cent of the world's evolutionary or genetic variation is present here. As it would be in most floras.

Of more note, around 40 per cent of Australia's native species belong to just five families.[16] Four of these families give the Australian flora much of its distinctiveness: Myrtaceae (the myrtle family, including the eucalypts), Proteaceae (the protea family, including banksias and grevilleas), Fabaceae (the pea family, including wattles) and Poaceae (the grasses, including spinifex, *Triodia*, which dominates much of arid Australia).[17]

The fifth family, Asteraceae (the daisies), also contains lots of endemic diversity but because it is the largest family worldwide and common in nearly all plant-inhabited places it doesn't scream 'Australia!' the way a gum tree or banksia does. Today there are very few parts of Australia – rainforests being a notable case – where you won't find these four, or five, families dominating the floral landscape.

Sixty-five million years ago, the flora was rather different. Australia was still attached by land to Antarctica, and via Antarctica to South America, having lost its connection to Africa and India around 50 million years earlier. This wet, seasonally dark remnant of Gondwana sat close to the South Pole and its forests contained conifers, ferns and ginkgos, along with a few familiar flowering plants such as ancestors of she-oaks (*Casuarina* and *Allocasuarina*), southern beeches (*Nothofagus*) and members of the Proteaceae.[18] All these plants are represented in Australia's flora today but only Proteaceae has flourished in terms of species (there are lots of ferns in many families, and none of them as diverse as the big flowering plant families mentioned above).

Based on recent DNA studies,[19] we can attribute the flora we see today in Australia to many causes, beginning with the slow movement over 3.4 billion years of large tectonic plates that contain today's continents and surrounding ocean. Add to that fire, global cooling events, weathering of rocks and soils, novel habitats, evolutionary innovations and, importantly, dispersal across oceans. While these forces have been active over millions of years, some have continued to shape the composition of Australia's flora since human arrival, 60,000 or so years ago, such as cultural burning, for starters, but also long-distance dispersal to and across the country.[20]

The big moves, though, were made millennia ago. Roughly half the 'lineages' (plants arising from a single ancestor) seem to have persisted in Australia since it was separated fully from Gondwana,

33 million years ago. The other (roughly) half are most likely to have travelled across the oceans since that separation, with a few lineages of indeterminate origin.

Families such as Proteaceae include a mix of origins – some persisting in the Australian part of Gondwana, others crossing the ocean since the separation of Africa.[21] Palms and many tropical rainforest species will be immigrants from northern lands that were never part of Gondwana.

Although a vast number of Australia's species probably arrived *after* the splitting of Gondwana, the relative isolation of the continent will have led to the arrival and establishment, as Australian biologists Mike Crisp and Lyn Cook put it, of a 'biased sample of the world's biota'. Once established these plants 'lucky' enough to reach Australia could evolve to occupy most of the available niches to the exclusion of other later arrivals. First in, best dressed, you might say.

That would explain why Australia wattles (*Acacia*), should they not be Gondwanic in origin, may have spread into and dominated drier forests and woodlands, joining eucalypts that were also capable of adapting to fire. Being a nitrogen-fixing legume in nutrient-poor soils wouldn't do any harm either, say Crisp and Cook. They conclude that 'the Australian flora is mostly a modified sample of the world flora, with some distinctive characteristics resulting from idiosyncratic events … early in its existence as an island'.

Australia's flora today may be in large part a lucky cohabitation, but it remains more than a little curious and distinctive.

April 2024

Exotic plants face a prickly reception

Plant vandalism displeasure rather than protest

We can't be sure who macheted the cactus collection at Melbourne's Royal Botanic Gardens on that Tuesday night in early winter 2013, or

why, but one motive postulated was a hatred of exotic plants – there are no cacti native to Australia.

The frenzied attack on all the tall, upright succulents occurred in a week that was not unusual back then, where Australians were having their attitude to Indigenous Australians and overseas asylum seekers tested almost daily. At the core of all these issues was a fundamental question about who can call Australia home and how we deal with the answer.

In one of his *Encounter* essays,[22] Milan Kundera quotes Vera Linhartova, a Czech author who like himself moved to Paris and from there wrote in French, as saying 'the writer is not a prisoner of any one language'. A great liberating sentence, says Kundera, and only the brevity of life keeps a writer from drawing all the conclusions from this invitation to freedom.

Lovely lines and part of a plea for writers in exile to be considered neither of their home or adopted country, but what he calls 'elsewhere'.

It seems harder for plants to be from elsewhere. They grow either in their home habitat, mollycoddled in an adopted garden or invade another plant's home to become what we call naturalised – established and self-maintaining in natural settings. Kundera was writing about how difficult it can be, for some, to accept a writer from no particular place. Can we accept a plant under the same conditions?

Living in Britain for 2 years, I came to realise that all plants there are from elsewhere. Some are arguably more native than others, but the immigrants have blended with the indigenous now and it doesn't make much sense to talk about natural and human-made landscapes.

Back in Australia, place of origin is of utmost importance, whether relating to Indigenous people, boat people or sports people. To Australians, it seems, it still matters.

That's not a bad thing when it comes to nature. We still have places some call wilderness[23] and we still have vegetation that contains almost the same suite of species as when Arthur Phillip arrived with the first European settlers in 1788.

Outside our more or less natural lands, in places where human impacts are stronger, the distinction becomes blurrier. It seems odd to build roads and paths, close-packed homes, car parks – indeed 'pave paradise' – but then worry about whether a tree in our garden is from home or away.

I've banged on about this for many years now,[24] but if a plant stays put (it doesn't spread into nearby bushland), doesn't need excessive water or nasty chemicals to survive, and is not harmful (and perhaps even attractive) to local wildlife (in the broadest sense), then plant it.

The pepper tree is a good example. There are some wonderful old specimens of *Schinus molle* in my neighbourhood. They conjure up vivid memories of my childhood in country Victoria, where they were the tree of choice around late 19th-century farmhouses, usually outliving the home.

The home of the pepper tree is South America, in the deserts and dry lands of Chile, Argentina and Peru. It had assisted passage to much of the Southern Hemisphere and in places it is a troublesome weed. In South Africa, for example, it is naturalised and causing widespread environmental damage.

In Australia the story is less clear. In some places it invades and displaces. In others it just persists, a reminder of past dreams and achievements in rural Australia, or here in the suburbs of Melbourne evidence of earlier planting fashions. In my mind, it comes from elsewhere, in time and place.

There are many other plants in the elsewhere category. Whether you choose to accept them or not will depend on personal whim and circumstance. I'd nominate oaks, elms, cedars, the jacaranda, a few scrappy weeds in our paths and more controversially, perhaps, the red flowering gum from Western Australia and Moreton Bay fig from subtropical and tropical eastern Australia. I'd also include cacti.

I suspect the rampage in the Botanic Gardens was a random act of vandalism, but should there be any link to the origin and worthiness of these plants, I would be bitterly disappointed. No plant should be the prisoner of a country or state.

The Age, June 2013[25]

Postscript

While no-one was prosecuted for this act of vandalism, or other related acts such as ring-barking the so-called Separation Tree, a river red gum under which Lieutenant Charles Latrobe announced in 1851 the impending separation of Victoria from New South Wales, we understand the perpetrator was more likely to have been unhappy with Royal Botanic Gardens Victoria over some minor operational matters than making a statement about the origin of species.

Still, as I return to elsewhere in this chapter and the next, plants from elsewhere are still considered by some to be less worthy than those from the political construct of Australia. Those with a more welcoming attitude to foreign plants sometimes extend their affection to weedy species. This may be based on the potential economic value of such plants – as argued by Nimal Chandrasena in *The Virtuous Weed*[26] – or through a mission to redress the extent of the 'problem' – as John Dwyer encourages us to do in *Weeding Between the Lines*.[27]

Dwyer reminds us that most plants from elsewhere do *not* become problem weeds. He cites 'Williamson's tens rule', which states that on average 10 per cent of introduced species will end up 'in the wild'; 10 per cent of these will become established; and then only 10 per cent of these become what we might call pests. That is, one in a thousand imported plants will become problematic.

Many would disagree with this analysis, and Dwyer mentions one example, Rochard Groves, the Australian weed-ecologist. Groves accepts Williamson's 10 per cent figure for the proportion of plant species likely to colonise a new environment but estimates 50 per cent will become established, with one or two becoming serious weeds. So, one or two in a hundred.

Either way, most introduced plants won't become weeds, which is Dwyer's main point. In Australia, of 26,000 introductions in the last few centuries, some 2,600 have become naturalised,[28] which amounts to over 10 per cent of the plants you'll find in the Australian bush. In the 'Garden State', Victoria, 28 per cent of the vascular flora are naturalised plants,[29] but that includes escaped crop and pasture weeds as well.

Even, as has been estimated,[30] plant species become naturalised at a rate of about 20 per year (the same rate, more or less, over the last two centuries), the vast majority of garden plants haven't escaped from the garden (yet) or if they have, they are not (yet) established and spreading in bushland.

You may not share John Dwyer's enthusiasm for weeds but as I conclude in my Foreword to *Weeding Between the Lines*: 'Like everyone, I love a book that affirms my biases and obsessions, but I much prefer one that does enough of that to allow me to trust the author, then prises open some of those cherished views for me to reexamine in a fresh light.' I hope to have done a little of that here, and elsewhere, in this book.

First Australian plants

What is a native Australian plant?

Simple, right? If a plant grows *naturally* in Australia, we call it a *native Australian plant*. If it grows naturally in a particular part of Australia, we might narrow down this definition and call it a species *indigenous* to that area. Otherwise, the plant is an *alien*, *exogenous* or perhaps a little less judgementally, *exotic*.

Take the Cootamundra wattle. It is a native Australian plant, indigenous to southern inland New South Wales, in the vicinity of a town perhaps better known as the birthplace of that exceptional cricketer, Donald Bradman. When planted or growing elsewhere in Australia, the Cootamundra wattle is exotic, but still a native Australian plant. The widely planted jacaranda is a native South American plant and exotic anywhere in Australia. So far so good.

But what do we mean by *naturally* or *native*? Given that over millions of years all plants have moved around the globe – evolving and annexing new territory across the now-disintegrated continents of Pangea and Gondwana perhaps, travelling across oceans in other cases – we don't mean they must remain fixed to one place. What we usually mean is that a species has ended up where it is, without obvious intervention by humans. Simple enough in the above examples but how direct or deliberate does that invention have to be?

If humans change the climate or move plant-distributing animals from one place to another, is that *unnatural*? What about interventions by Aboriginal people who have inhabited a place like Australia for at least

60,000 years?[31] Perhaps the biggest question of all, why do we separate the influence of one species – *Homo sapiens* – from others; are we, as a species, *unnatural*?

Putting aside the *naturalness* of humans, there is an equally complex nomenclature to describe the different ways a plant ends up back in *nature* after it has been carried to a new location via a garden or paddock. We often talk about a plant *escaping* into *nature* and becoming *naturalised*, when it is sourced from a place tended by humans. Victoria's online guide to its *native* and *naturalised* (from outside Victoria and now freely reproducing and well established in the State) plants, *Vicflora*,[32] recognises four different *degrees of establishment*: *native*, *established*, *casual* and *reproducing*. The last three categories are for plants that were not present before European humans first settled in Victoria.

In my home state, Victoria, *European disruption* to the lives of resident people and plants began in the 1830s. Some of the plants brought into the State, deliberately or inadvertently, will have already become *naturalised* at earlier settlements such as those established in 1788 around Sydney Harbour. As to the status of plants growing anywhere in Australia before European settlement, the simple answer is that the plant is *indigenous*.

While that feels right for plants that have evolved on this land over millions of years, it sidesteps the problem of any plants brought to Australia by traders visiting our northern shores before 1788 or, even more problematically, those brought with or further aided in their dispersal[33] by Aboriginal people who arrived here around 60,000 years ago. Should such plants be considered indigenous or alien? Most people, or at least most botanists, would not consider a plant indigenous unless it had spent a few thousand years here at least. But where do you draw the line and, perhaps more interestingly, why?

In 2007, Australian botanist Tony Bean reviewed[34] how the term *indigenous plant* is used outside Australia. In Europe, it's a plant that has been present (or known to be present) since the beginning of the Neolithic period, 12,000 years ago, or that has arrived since then unaided by '*human activity*'. There is often a further distinction made between plants introduced before 1500 AD (*archaeophyte*) – up to and including the Middle Ages – or after (*neophyte*).

That 1500 AD date has been transported to North America, but for different reasons. It is close to the arrival of Christopher Columbus at the end of the 15th century, after which European colonisation led to the same devastating transformation of the country's landscape as occurred in Australia post-1788. For this reason, in North America, plants introduced and moved around through human activities after 1500 are considered *alien*.

In New Zealand, the arrival of Polynesian settlers, the Maori, 1,000 years ago, is used as an earlier starting (or is it finishing?) point for that country. The first European contact, in the late 18th century, simply amplified the opportunities for human interference in plant distributions. To generalise across New Zealand, Europe and North America, humans seem to be considered separate from nature when their impacts are judged to be *destructively transformative*, due to population numbers and/or mode of living. Not an unreasonable perspective, but this definition is fraught.

To transfer such concepts to Australia, we might consider as *alien* anything introduced by human activity since (1) 1788, with European settlement, (2) 5,000 years ago, when we have the oldest archaeological evidence of Asian traders visiting Australia or (3) 60,000–70,000 years ago when the First Australians arrived. (1) would follow the North American precedent, (2) the New Zealand one perhaps and (3) the European. Or we could follow South Africa's example, which has no time-limited definition – a plant whose introduction can be linked to human activity is considered alien.

In any case, can we determine if an apparently alien plant was introduced 230, 5,000 or 60,000 years ago? It is hard enough for some plants to decide if they are part of some ancient distribution, perhaps including long-distance dispersal by animals or other vectors, or whether humans are somehow implicated in their present-day placement. One extreme view cited by Bean, is that 'where there is an apparent disjunction overseas [Bean has arbitrarily set this at anything over a 2,500 km gap] at the species level, there has either been a misidentification, or the plant is introduced'.

Bean argues for at least three groups of plants where this is patently not so – coastal plants such a mangroves, inland aquatics and plants with adhesive fruits or seeds – and adds additional caveats to the definition of apparently indigenous for the creation of his diagnostic key to alien versus indigenous plants.

In some ways it can be easier to simply set dates and ignore the practical aspects of how they got where they are. For example, if a plant was reported by the first European botanists, then we consider it *indigenous*, whether or not it came in the week before or a few thousand years earlier.

This isn't very satisfying for either scientist or layperson. I think most people would consider a plant that fell from a wheat sack on the way to Parramatta as somehow different in origin to a species that evolved as part of the great Australian flora inherited mostly from Gondwana and our near Asian neighbours – what we might call our *First Australian plants*. Then there is the money to be spend on conserving and saving species or eradicating the weedy ones that might displace native ones or reduce crop yields.

Such a decision is sometimes arbitrary, and we might be best making a call based on the best environmental and financial outcome, no matter what we think or know about the origin of the species. That leaves the categorisation of native or not unresolved but for this purpose at least, irrelevant.

April 2024

Browned off

Keep the lawn going or let it die?

If you like brown grass and a scraggy gum tree or two, you are in luck. Your time has come. As we navigate our way through water-challenged times, a parched scrap of native grassland and a few remnant eucalypts will provide a guilt-free garden.

On the other hand, if you favour exquisitely manicured green lawns with a lush border of annuals, be prepared to be marginalised by your gardening comrades as excessively extravagant and wilfully indifferent to the plight of our planet.

At least that's one perspective. As with much in this world, such a didactic response is inadequate for what is a complicated and deeply nuanced issue. I think you *can* have your green lawn and enjoy it too.

If you don't mind a bit of summer brown-off, you can have what I like to call the olive-green lawn. Green in winter, brown in summer – combine the two colours and you get something like olive-green.

In much of southern Australia, the olive-green lawn will give you up to 10 months each year of greenness, without any additional water,

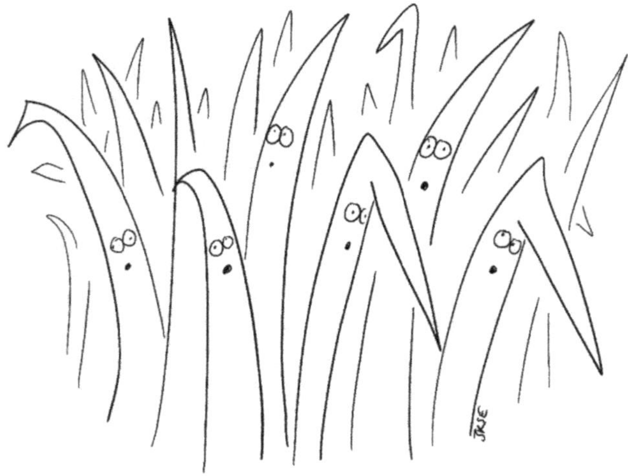

fertiliser or toxic chemicals. Your environmental footprint will be shaped more by your mower fuel: electricity, petrol, hand or perhaps rabbits (as has been suggested by author and gently provocative garden writer, Jackie French).

It is true there is generally little biodiversity associated with your typical lawn, so letting the grass grow a little longer and allowing a few adventives to take root, would be a good thing. As would keeping the lawn rather small in area and enclosed within richly planted garden beds.

If you do decide to irrigate over summer, perhaps from tank or recycled water, there are some extra benefits. Lawns keep us cool. Grasses, like all plants, absorb air pollutants and noise. And we feel happy and relaxed on a soft, green expanse of grass.

Compared to many garden plants – and home vegetables fall into this category – a lawn needs less water to survive, and even flourish. You can also water your garden indirectly by watering the lawn, assuming the latter is relatively small compared to the former.

You might also consider a native grass lawn, which tends to require less water and fertiliser. Interplant with local wildflowers and you'll encourage even more wildlife. Don't mow too often or too low so that whatever grass you chose will need less water.

Mostly though, don't be ambivalent about your lawn. Think it through then make it work for you and the environment. The same goes for the garden you create around that lawn. You might hanker for a border of lush, soft plants from the wetter parts of Australia or overseas. Perhaps annuals are your thing.

Again, be confident and considered. Gather water-hungry plants together and if you can, collect rainwater or reuse greywater from indoors. You might even wish to wash yourself and your clothes less and use your 'water allocation' for the garden, although this may require some negotiation with your local water-issuing authority.

All this will be particularly important if you have seasonal or permanent water restrictions in place. These are typically a very blunt instrument, but understandable given the complexity in enforcing them. While constraining outdoor use before indoor might work for many, it won't suit the keen gardener – who, incidentally, might be willing to pay for the privilege to use more water on their garden. Funds which could be returned to offset other water-saving initiatives such as tree planting or (more controversially perhaps) a sustainably powered desalination plant.

For now though, consider a watering system to give an even and more disciplined spread of water to the garden or the lawn. Drip irrigation gets water closer to the plant roots and there is less evaporation and run-off, but it's slow, outlets can clog easily and in some soils the water may not soak through to where plants need it most.

If, when the calculations are done and your priorities weighed up, you decide to grow more drought-tolerant plants and decommission your lawn (or let it brown off completely in summer), what would you like to see in your local parks and gardens?

They can either follow suit, sharing experiences of how to garden in a water stressed world, or they might provide oases of green for all the community to enjoy. Some places, like botanic gardens, also hold valuable collections of plants for conservation, science and education, which complicates that decision.

Generally, I find people support keeping their parks inviting and green, and their botanic gardens alive and functioning, even under the most extreme conditions. The City of Adelaide, for one, publicly supports the watering of public lawns and gardens in summer, to keep that city cool.

Keeping public spaces green need not come at a high environmental cost either. Irrigating all the lawns and gardens in Royal Botanic Gardens Melbourne took on average a cup of water per visitor per day, and where possible that water was from sustainable, non-potable sources.

The benefits are tangible. Public green lawns can be a salve on a cripplingly hot day in summer, a place to feel – and to be – calmer and cooler. On a 30°C day in Melbourne, the temperature just above the lawns in the city botanic garden is about 24°C. The nearby bitumen paths are a sizzling 40°C (over 100°F).

A combination of large shady trees and lawn is even better. The Oak Lawn in Royal Botanic Gardens Melbourne is as cool in summer as the nearby Fern Gully, with its running water and lush vegetation. Both are up to 6°C cooler than the surrounding city when temperatures elsewhere rise above 30°C.

No matter where you live or work, the cost benefit of a lawn is worth calculating. The familiar turf roof on Canberra's Parliament House can be 22°C cooler than nearby granite paths, which helps to keep that building cool. It's a cheap shot, but that might just make up for all the hot air inside.

Gardening Australia, February 2020[35]

Postscript

I wrote a similar piece for the ABC website in 2017,[36] following a radio chat with Jonathan Green on ABC RN's *Blueprint for Living* about the merits and demerits of lawns. While I covered some of the same turf (sorry), I began with a quotation from author Yuval Noah Harari.

In his thought-provoking 2016 book, *Homo Deus: A Brief History of Tomorrow*,[37] Yuval Harari described lawns as a petit-bourgeois symbol of 'political power, social status and economic wealth'. By not establishing a lawn in your new home, he argued, you can 'shake off the cultural cargo bequeathed to you by European dukes, capitalist moguls and the Simpsons'.

I went on to cite the Permaculture Research Institute in northern New South Wales,[38] which began an encyclical on lawns with a similar revolutionary cry, reminding us that the origin of lawn growing was with the 'fabulously wealthy'. Perhaps worse still, the English wealthy.

After citing the many (true) undesirable impacts of a lawn – fewer plant and animal species, wasting water and the need to employ 'weapons against nature' (think lawnmowers, trimmers and chemical sprays) – we are told the proletariat was 'rob[bed] ... of at least one of the two days of freedom and leisure' each week.

Against this this backdrop, I made the same, hopefully nuanced, case for lawns as I did in *Gardening Australia* magazine. I also included design arguments – creating a 'void' within the mass of shrubbery – and their ability to absorb air pollutants and noise. I quoted experts as saying you can have a green lawn in Melbourne for 10 to 11 months a year without irrigation, and there is no need to fertilise turf 'unless you have heavy foot traffic or want to keep your lawn particularly swanky'.

For good measure, I added a provocation put to me by the senior curator of horticulture at Royal Botanic Gardens Melbourne at the time, Peter Symes. He said if the Gardens wanted to save water, we would remove all our trees and garden beds, and replace them entirely with (preferably warm season) lawns.

I ended my deliberations by saying the jury was still out on the merits of a (mostly) green lawn: 'Perhaps reaching its verdict on a shaded lawn near the courthouse. If that lawn is mown high, unwatered, unfertilised and contains grasses suited to the local climate and soils, then the 12 good men and women are in the perfect place to decide.'

As for my own home. We have a small front and back garden, both with lots of plant variety (plants from Australia in front, mock-subtropical at the back) and no lawn. We've just converted our nature strip into a mix of local grasses and grassland species, with a few other things we like from elsewhere in Australia.

Bad blood

> *Like an egg stain on your chin, you can lick it, but it still won't go away*[39]

The floating islands in the Ornamental Lake at Royal Botanic Gardens Melbourne are at their prettiest in late summer when the purple loosestrife is in full bloom. Purple loosestrife is a rugged and

serviceable plant, with willowy leaves and showy purple flowers. It favours swampy ground and lakeside settings.

Perfect for this situation, tarting up flotillas made from recycled plastic bottles. Well, apart from some uncertainty about its origins and intent. Most of the other plants on these islands are Australian natives and this prolific flowerer looks weedy. In addition to its brightly coloured flowers, which wouldn't be out of place in an English perennial border, it is disconcertingly resilient.

Five of the 36 species of *Lythrum*, the loosestrifes, grow wild in Australia. Two of these, *Lythrum wilsonii* and *Lythrum paradoxum*, are endemic (found only in Australia) and restricted mostly to the dry interior of the continent. The three others are 'probably native'.[40]

Lythrum hyssopifolia is considered native to the Mediterranean and 'apparently native' to eastern Australia but naturalised in Western Australia, where it is widespread in the south-west. *Lythrum junceum* is another Mediterranean native considered a more recent immigrant to Australia – so perhaps miscategorised as even apparently native.

The topic of this essay, purple loosestrife, *Lythrum salicaria*, is usually described as 'cosmopolitan'.

The botanical name *Lythrum salicaria* is from the Ancient Greek word for black blood or gore, *lythron* – and because no flowers in the genus have such sombre coloured flowers this probably[41] refers to the reputed blood-clotting uses of the plant – and *salicaria*, which means willow-like, which it is in leaf and to some degree flower arrangement.

The source of the common name, loosestrife, is most likely to be a misinterpretation of the botanical name given to an unrelated group of plants, *Lysimachia*, also commonly called loosestrife.[42] That genus name has been mistakenly thought to be adapted from the Greek word *lusimakhos*, meaning 'ending strife' (literally 'loosening battle'). In fact, it honours the rather notorious Lysimachus, who became King of Thrace, Asia Minor and Macedon in 306 BCE, following Alexander the Great. At least the common descriptor for our species, 'purple', is literal and correct for the flower colour.

Whatever you call it, purple loosestrife does well in the damper parts of south-eastern Australia. It also flourishes in the marshes of New England in the United States: 'choked' is the word used by Anne Raver in a recent *Landscape Architecture Magazine* article about New York State's plan to curb the spread of invasive species.[43] If fact it is now illegal to sell *Lythrum salicaria*, along with 68 other 'banned, non-native, invasive' species in that part of the United States.

Purple loosestrife is described on the US website Invasiveplants.net[44] as:

> a plant of European origin [which] has spread and degraded temperate North American wetlands since the early 19th century. The plant was introduced inadvertently as a contaminant of European ship ballast and deliberately as a medicinal herb for the treatment of diarrhea, dysentery, bleeding, wounds, ulcers and sores.

It was also 'bred and planted for horticultural purposes' in North America and has clearly been introduced 'multiple times'.[45]

Transporting myself back to Kew Gardens in London, where I worked for 2 years in the early 2010s, I recall seeing purple loosestrife in many places beside the River Thames. Every week I walked and cycled beside the Thames, kayaked on it (once passing the Royal Barge as it made its way back upstream to its home somewhere near Teddington Lock), and lived less than 100 metres from its banks. Purple loosestrife was just one of a seasonal parade of colourful riparian wildflowers.

According to Tom Cope in his 2009 book *The Wild Flora of Kew Gardens: A Cumulative Checklist from 1759*, purple loosestrife is native on the towpath but planted and grown since 1768 inside the Royal Botanic Gardens beside the lake. My office at Kew Gardens was on the first floor of Museum No. 1, and from my window I looked across the lake towards the better-known Palm House designed by Decimus Burton (with Richard Turner) and shaped like the upside-down hull of ship.

Each weekday morning for those 2 years I walked from my home near Kew Palace to my office, around the lake, passing both the Palm House and seasonally flowering purple loosestrife.

While the English claim this species as a native, as in much of that country's flora, this language is imprecise and may mean only that it has been growing somewhere on the continent and nearby islands since the time of King Lysimachus (or more technically, further back to the neolithic).[46]

Despite its appearance, purple loosestrife is considered *native* in most of Australia, meaning it is assumed to have been growing here before the First Fleet arrived in Port Jackson in 1788, and presumably before it was planted in Kew Gardens. As you'll recall from the first essay in this chapter, this doesn't imply it was 'always here'.

Still, it's an unusual distribution. Very few plants are truly cosmopolitan (found on most continents, let's say) unless they are considered 'weeds' for at least part of their world distribution. In Australia, purple loosestrife is mostly a plant of the south-east,[47] with a cluster of records near Adelaide and eastern Tasmania, then a broad sweep from the far eastern corner of South Australia through to Brisbane and thereabouts.

According to the *Atlas of Living Australia*,[48] purple loosestrife is 'near threatened [with extinction]' in South Australia and 'endangered [close to extinction]' in Tasmania, and native but under not risk of extinction in Victoria, New South Wales and southern Queensland. Yet despite all this, purple loosestrife was included in a 2003 list of weeds in Australia,[49] albeit with the caveat that the species is native in *some* States.

While it is possible for a plant to be invasive in some settings, and some countries, and not in others, this is again unusual for a plant of such obvious virility and one that inhabits swamps and inundated land. Seeds and plant fragments of such species can be distributed locally by flooding waters, and between countries by migrating aquatic birdlife.

If we look to international sources, the Commonwealth Agricultural Bureaux International's – aka CABI's – Invasive Species Compendium,[50] lists purple loosestrife as an 'Old World native', with a natural distribution throughout Europe (except for high mountains and the far north), extending westward into Russia, Iraq, Iran, Afghanistan, China and presumably neighbouring countries across the continent of Eurasia, *and* Australia. There are also other reports of it growing naturally in Japan and northern India.[51]

In South America, there is a single record from Argentina, as an introduced exotic. New Zealand records it as a non-native weed. The South African National Biodiversity Institute considers purple loosestrife to be native to Europe, Asia, Australia and northern Africa (you can find reports from the northern parts of Morocco, Algiers, Tunisia, Libya and Egypt).[52] The Institute categorises the species as invasive in south and west Africa, as well as in North America. It is uncommon in South Africa, seemingly confirmed to be a garden escape only near Liesbeeck River in Cape Town.

On face value then, we have a species with an apparently natural distribution sweeping across temperate Europe and Asia and then jumping across to Australia, but quite capable of establishing itself easily in new territory such as North America and Africa with a little help from humans (the species was accidentally introduced into North America around 1800).

Local variation in genetics or form can reveal origins and distribution paths, and *Lythrum salicaria* has been subdivided by taxonomists to account for variations primarily in hairiness and (in a somewhat circular way for my purposes) geography.[53] The Australian entity has not been given any separate taxonomic recognition, although, as I'll get to, it has some affinities with Japanese populations.

Australian botanist Tony Bean (whom we met in an earlier essay)[54] reminds us that while purple loosestrife is often described as cosmopolitan, it is accepted by all as introduced and 'an aggressive weed' in North America.[55] It is also absent from South America, apart from what is considered a minor weedy incursion. Elsewhere, the species is, he notes, generally considered non-invasive.

Bean's attempts to redefine our definitions for indigenous and alien (or exotic) plants has relevance here. His view is that a species should be considered *indigenous* if it meets the following criteria: there is good evidence for it being in Australia prior to European settlement (1788, at least around Sydney); it frequents habitat largely unmodified by those Europeans; it is not particularly invasive or showing evidence of expanding its range; it gets frequented by pests and diseases; it has some diversity in habit and genotype; and any discontinuities in distribution relate to non-human influences (although presumably not including here major discontinuities due to land clearing and the like separating populations since European settlement).

In some cases, one or more of these 'ecological criteria' may not be met, and that's fine for Bean, if the other evidence is strong. Bean also separates out special categories of plants such as coastal plants, aquatics or semi-aquatics (which could include our purple loosestrife) and those with 'adhesive fruits', which are more easily distributed over long distances by birds and other non-human animals.

You might expect such plants to be spread easily between continents even without human intervention. In the Australian region that intervention could date back thousands of years, with maritime traders and explorers visiting the region over at least the last 3,000 years – bringing domesticated animals and well-documented and deliberate plant introductions such as coconuts, foxtail millet and bottle gourd – and the Aboriginal peoples arriving here around 60,000 years ago.

Bean identified other supporting criteria, such as the species having relatives growing in Australia, being present (and apparently indigenous) in places close to Australia, and/or it not being recorded as an invasive or naturalised plant elsewhere in the world. Bean does make the point that all this is complicated by the fact that a species can be indigenous in one region or habitat and invasive or alien in another. In theory this could be within a country, State or even smaller region.

In this context, Bean considers purple loosestrife *not* to be an invasive species in Australia. He acknowledges that there are differing views on its status, but his starting point is a collection of the species by the visiting botanist Robert Brown at Port Dalrymple, in Tasmania, in January 1804, and soon after in Port Jackson, New South Wales.

These two gatherings are within the first few decades of European settlement, although there were a few shiploads of people, and no doubt plants deliberate or otherwise, entering the country during that time. Later in the century,[56] the Victoria's first Government Botanist and director of the Royal Botanic Gardens Melbourne, Ferdinand von Mueller, considered it native. The explorer Thomas Mitchell found it abundant in the Macquarie River area and the Reverend William

Woolls had seen collections of it from Mudgee and elsewhere in New South Wales.

Further support comes from the number of chromosomes found in plants from Australia and Japan, which are different to those in Europe, suggesting a separation of these populations from those of Europe for a period longer than a few hundred years. The clincher, for Bean, is the presence of purple loosestrife pollen from prior to European settlement in cores extracted from Wingecarribee Swamp, just south of Sydney.[57] The species, he concludes, 'occurs in intact vegetation communities and its geographic range is seemingly unchanged over the past 100 years'.

Against all that is the evidence of strong invasiveness in North America and the rather odd disjunction of its distribution from the Northern Hemisphere to the south-east of Australia.

On balance though, I'm with Bean, and (I think) the precautionary principle would have us act as if purple loosestrife is indigenous to Australia. Similar arguments have been made for the water primrose, *Ludwigia peploides* subspecies *montevidensis*, which was collected around Sydney in 1803 and 1804.[58] In Victoria it is considered 'most likely' to be indigenous except in disjunct occurrences around Melbourne.

Neither species is spreading aggressively so even if we are wrong, it probably doesn't matter. What is important is that we consider this and other cases on their individual merit. Bean assesses 39 other species of dubious indigeneity, deciding 30 are aliens, two of mixed origin, and only seven should join purple loosestrife as indigenous.

Back in New York State, some invasives, like purple loosestrife, are banned while other, lesser evils are to be regulated but not prohibited. For the likes of burning bush (*Euonymus alatus*) and Norway maple (*Acer plantanoides*), gardeners and nurseries must tag the plant with a label that specifies what to do to stop it spreading (e.g. deadhead before seeding) and suggesting alternatives.

The New York scheme includes a formal process to review determinations and to tweak the listings. For example, if some cultivars can be shown to be less invasive than others, then they may be delisted. Prosecuting these cases will be fascinating to watch and may well have implications for how we treat our pretty purple loosestrife.

April 2024

Boab dreaming

People and plants with a shared history

The very best definitions of a *native* or *indigenous* plant, according to our now much-cited Brisbane botanist Tony Bean,[59] combine elements of longevity ('in that region for thousands of years'), absence of human

assistance ('not brought there by direct or indirect human activity', 'arrived there without intentional or unintentional intervention of humans') and originating from another native population (if transported by non-human means, also 'from an area in which they are native').

As to that final point, Bean wants to exclude the obvious loophole where a weed from somewhere else in Australia is transported to a new place by any non-human. This does not, says Bean, make it indigenous in its new location. I think we'd all agree with that, even if an indigenous animal does the heavy lifting. A blackberry seed ingested and disgorged by an emu doesn't become transformed into a native species by a trip through a native animal's digestive tract.

What about a native plant, transported by an alien animal? Things get messier then. A blackbird distributing seed of local wattle might establish a weedy population of an otherwise indigenous species. And what is that alien animal is a human? Anything post-1788 we might consider the equivalent of a blackbird, but what about plant movements aided by First Australians? Recent research on the boab has created quite a quandary for anyone wanting the see the world of native and alien plants in black and white.[60]

The boab is an odd-looking tree wherever it grows, with a bottle-like trunk sometimes looking like it has been plucked from the ground and then replanted upside down. It has also been compared with a giant stick of celery. Today, the boabs, a collective common name for species of *Adansonia*, grow *naturally* in Africa, Madagascar, the north-west corner of Australia and across Asia into Malaysia.

Six of the nine species are *native* to Madagascar, and two to the continent of Africa. One of the African species, *Adansonia digitata*, was introduced by humans into Madagascar as then transported through to Malaysia, and is now considered *naturalised* in both places.

The ninth species, *Adansonia gregorii*, occurs only in the Kimberly, in northern Western Australia, and a little further east in the Victoria River district in Northern Territory. It is considered, quite reasonably, to be an Australian *native* plant with obvious historical connections to

Africa and Madagascar, connections that reach all the way back to the Great Southern Land, Gondwana.

Or perhaps not. As you would expect for such *charismatic megaflora*, the ecology, biology and taxonomy of the boabs has been reasonably well-studied over the 250 or so years since its scientific naming.[61] That said, it wasn't until 2018 that a second African species, superficially like the first, was confirmed by the number of chromosomes and a comparison of the genes carried on them.[62] The advent of sophisticated molecular sequencing has also dampened enthusiasm for an ancient Gondwanic origin of Australian species.

As with all organisms on Earth, the evolutionary history of the boab is etched in their genes: by looking at shared and conserved regions of the genome, we can retrace historical relationships between all species on Earth today. From such comparisons, we now know the isolation of the trees in northern Australia from those in Africa and Madagascar occurred just over 6 million years ago. Well after the split of western (Africa and Madagascar) and eastern (Australia) Gondwana, about 180 million years ago, but well before the arrival of the First Australians, around 60,000 years ago.

For part of those 6 million years, the Australian cohort may have settled somewhere 'offshore' – perhaps on land bordering the Indian Ocean – and there became partly or fully established as a distinct species over the few million years required for this transformation. Then on to Australia through whatever means.

Retired Queensland neuroscientist Jack Pettigrew is one who favours this theory, with the final journey to Australia aided by First Nations people or others arriving on Australia's northern shores. Pettigrew argues that because Africans value the fruit as food and regularly carry seed with them, there is a ready vector at hand to travel between trading regions; a fascinating but unsupported idea, with no evidence in boab genes or fossils.

The most likely way the boab got to Australia (or to Pettigrew's transit land(s) for that matter) was through saltwater-resistant seed of

Adansonia digitata drifting across the Atlantic, perhaps attached to other debris. Given a few million years, it is not unreasonable that at least one seed made it across the ocean alive. It's unlikely this will ever be proved, or disproved, but it's a satisfactory assumption.

That, however, is only part of the story. Once in Australia, how did it get about? Plants and animals occurring naturally in the Kimberley develop genetically distinct groups after they become isolated during floods or long-distance animal transport. The boabs haven't done this. There is little genetic variation in the species, and none correlated with barriers that would stop flooding waters or non-human animals.

Human-assisted transport is the most likely answer. Seed and seed pods are found in ancient Aboriginal sites in the region and the fruit is used for food and celebration. To support this idea, Melbourne-based geographer Haripriya Rangan and colleagues looked to Africa where there is good evidence for people distributing *Adansonia digitata* around that continent and abroad.[63]

Languages also travel and change. As Rangan says, in a piece written for *The Conversation*,[64] 'when people carry things from one place to another, they also bring their words for these things'. New words are created or borrowed (and often modified) for new things.

The overlap in the Kimberly between the evolution of Aboriginal words for boab and gene flow within the species was enough for the researchers to be convinced that humans were the main agents of transport once the boab reached Australia. The species arrived in the far north-west of the region where, in concert with a changing climate, it spread – including into areas now under the sea – and contracted into suitable habitats near coastal north-western Australia.

After hundreds of thousands of years of interbreeding and mutating, these populations evolved into a new species, a species then distributed more widely after the arrival of the First Australians. Depending on how you view the interventions of the First Australians, that makes this species either partly or fully *native* across its range in Australia.

Now, for a moment, forget what I've told you about this recent scientific and ethnobotanical research. Picture yourself driving through the red sand of the eastern Kimberly, as I was a few years ago, and pulling up beside one of those ancient-looking boabs with a bulbous trunk nearly 5 metres in diameter. Like me, you might take a photograph first. Then you can step back to take in the grandeur, and weirdness, of what must surely be part of Australian deep evolutionary history.

If you are a disciple of Dr Bean, as I am, you might consider for a moment whether this odd-looking plant really does *belong* here. The tree looks like it was here long before European settlement, and it behaves like a local plant.[65] The species is known to be from any other part of the world and its closest relatives all occur in two other fragments of Gondwana, Africa and Madagascar.

The boab doesn't seem to have what we might call 'weedy tendances', such as being short-lived and quick to flower, bearing seeds that adhere to passing creatures, or favouring watery places where floods and tides might carry it over large distances. (I should note here that many authentically native plants have these traits and move around within their *natural* range through these contrivances.)

Even without the recent research, the boab doesn't meet a couple of Bean's tests of indigeneity. First, there are no related species also assumed to be native to Australia. That shouldn't concern us too much given that species can be lost through extinction – recently and in the past – and in any case, this is a relative clause, given there are degrees of relatedness between all plant species. The Australian boab also has discontinuities (and, for that matter, continuities) in its distribution, which if attributable to humans can be symptomatic of *alien* status.

In our naïve state next to that soon-to-be-Instagrammed tree, we might assume that any gaps in the distribution of this species would be due to soils, rainfall, competition with other plants, or a myriad of other influences right through to charming haphazardness or luck. In any case, Bean doesn't suggest native plants need to meet all his rules, only that we should test them against these criteria and see whether on balance a plant would be most logically considered a native or a weed.

So back in the land of knowing, do we treat some or all humans as external to the processes of plant evolution and distribution? Sometimes this seems reasonable, other times not. For the Australian boab, we have the option of considering it a mix of *indigenous* and *alien* populations (like the Cootamundra wattle, *Acacia baileyana*)[66] or, if we consider pre-European human dispersal in Australia as different to that of post-settlement distribution aided by humans, entirely *native* or *indigenous* throughout its range.

We need not panic about all this unless our decision leads to the eradication of the boab from all but coastal areas. Which would be a rather extreme and perverse reaction, I would have thought. Categories such as *native* and *weed* work reasonably well, in most cases. With the boab, its status is more to do with semantics and intellectual curiosity. On the other hand, it does beg again that nagging question about the degree to which humans – First Nations or more recent immigrants – are part of nature.

April 2024

Palm gods from the north

Traditional ecological knowledge yet again informs scientific study

Outspoken and typically insightful Australian ecologist, Professor David Bowman, announced recently he was amazed, but not surprised, to read of an Aboriginal story recorded in 1894 by German anthropologist and missionary Carl Strehlow.[67] The story tells of how 'gods from the north' brought palm seed to central Australia, to what is now called Palm Valley, in Finke Gorge National Park, about 120 kilometres south-west of Alice Springs.

Palm Valley – *Mpurlangkinya* in the local Arrernte language – is an unlikely place for palms to grow. I've never been there but from photographs I have conjured up a Saharan-like oasis (largely based on mental recreations from Hollywood film) surrounded by an equally cinematic backdrop of red central-Australian rock. An exotic place with an equally exotic, and unexpected, dominant palm.

The evocation of a god to transport this species south make sense given the next closest palms of any kind are more than 800 kilometres away, beside the steamy blue water of Mataranka Hot Springs and then a little further away to the north-east in Lawn Hill National Park, well into the tropics.

The Arrernte palm origin story was recorded in German by Strehlow, after being passed down from generation to generation for more than 7,000 years, possibly 30,000, in local Indigenous language.

It was translated into English only shortly before reaching Professor Bowman in 2015.

The palm itself had been known to Europeans for a little over 150 years, since dogged explorer of Central and Western Australia, Ernest Giles, observed and collected the plant in 1872. Six years later, Melbourne's Botanic Gardens director Ferdinand von Mueller, described the Fink River palms as a new species, *Livistona mariae*, distinct from those in the land of the gods.

The palm is also known as the Central Australian cabbage palm, or sometimes red cabbage palm after the distinctive colour of the seedlings. I can't find any record of what it was called in the local Arrernte language.

There has been some debate over recent years about just how distinct the palms of Palm Valley are from those in the north, with some preferring to treat them as a subspecies of what is commonly called the Mataranka palm. This demotion in taxonomic rank still implies the southern palms have been isolated long enough to interbreed only among themselves and to exhibit a range of small variations in the way they look.

In the National Recovery Plan for *Livistona mariae* subspecies *mariae*, as it was called in 2008, the palm was considered ('believed') to be 'a relic of a once mesic climate in central Australia', a survivor of the great drying out of Australia around 15 million years ago.[68] The annual rainfall where it lives today is around 25 centimetres a year, perhaps surprising for an area described as arid but far less rain than received in the palm habitats in the north and east of the country. The Central Australian cabbage palm now only grows in gorges with permanent water and where it is protected from fire.

From the 2008 report, we also learn that seeds from the palm are eaten by the western bowerbird and spiny-cheeked honeyeater, among other bird species, and that they are likely to help in the perpetuation of the species today. Mostly, though, it was assumed the species spread its seed through moving water and gravity (dropping from the tree), and that populations upstream didn't interact much with those downstream.

While the palm was said to be of 'significant cultural importance' to local Indigenous Australians, there is no further mention of their interaction with the species. Equally there is no mention by scientists of gods or god-like people intervening on the palm's behalf.

That is, until Professor Bowman and his colleagues dropped their equivalent of the god story in 2012.[69] Their comparison of DNA from palms growing in Palm Valley with those at the two closest northern locations, showing that the lineage had split within only the last million years. This is well after the drying out of Australia and implies a different kind of origin story to that reported in the earlier scientific literature.

The team also concluded that not only is the Central Australian cabbage palm not a distinct species but it should not even be afforded the rank of subspecies. Its arrival in the south is far too recent to allow the kind of changes we expect for a distinct taxonomic category, although they do admit that due to the palm's clear isolation from northern members of the species, they seem to be in the process of 'speciation'.

The new evidence does not support the fragmentation of a once more widespread species or the further migration of individuals after the original settling of Finke River. Given this scenario, they concluded that the plant was as likely – perhaps more so – to be transported by Aboriginal people as fruit-eating birds and bats.

Moreover, that seed was most likely to have been carried deliberately to these oases in Central Australia; not unlike the boab story,[70] but in this case with the disjunction on the same continent rather than separated by the Indian Ocean.

Even before reading the Strehlow translation in 2015,[71] it was clear to Professor Bowman and colleagues that the seeds must have been carried to the Central Desert up to 30,000 years ago. The Aboriginal legend of perhaps similar age was effectively telling the same story.

'The concordance of the findings of a scientific study and an ancient myth', said Bowman, 'is a striking example of how traditional ecological knowledge can inform and enhance scientific research.' Not only that, 'it suggests that Aboriginal oral traditions may have endured for up to 30,000 years, and lends further weight to the idea that some Aboriginal myths pertaining to gigantic animals may be authentic records of extinct megafauna'. And, we might add, of extinct and interesting megaflora.

April 2024

Handful of beans

The gods may have borne both palms, beans and much more

Here we go again, I thought, as I read the first reports of humans assisting the black bean, *Castanospermum australe*, to get to where it is today. Not as a feature tree in parks and gardens, but scattered through coastal rainforests of north-eastern Australia, where its present day 'natural distribution' is apparently thanks to the careful culinary skills of our First Australians.

I wondered, only half in jest, whether any plants in Australia are today found only where they were prior to the First Australians arriving. Maybe the European invasion and subsequent ransacking of the environment wasn't so different after all.

Of course, it *was*. The scale of change since 1788 has been, and is, entirely different. But the more we find out about the history of our flora, the more we need to change our view that the Australian landscape in 1788 was a thing free of human manipulation. The authors of the 2017 paper on the black bean were both deferential and a little dismissive of the boab and Central Australian cabbage palm research,[72] so I was keen to see what they had done differently.

According to Sydney conservation geneticist Maurizio Rosetto and his colleagues,[73] their study 'integrate[d] from the onset,

anthropological, molecular and ecological research to test if Aboriginal-mediated dispersal can explain the distribution of a valuable but non-cultivated resource tree'. The answer, as with the previous two non-cultivated resource trees (if you'll allow me to grant tree status to the palm based on common usage of the term), was 'yes'.

The black bean is found beside rivers and streams in old growth forest as well as newly disturbed gaps. It is a tall tree, up to 40 metres high, with glossy, fern-like leaves. In late spring (my 'sprummer'),[74] chunky orange-red flowers emerge from the stems and trunk, transforming through summer into long seedpods containing usually two to five, matte-brown seeds, each about 3 centimetres across. You can grow these seeds at home or purchase them as 'lucky beans' where the seed rests on the soil surface with its two halves (the cotyledons) prised apart by the emerging true leaves.

The toxins in the seeds are deadly unless used judiciously as a medicine for now identified anti-viral properties. Yet the seeds can be made edible with extensive pre-treatment, a process honed over 2,500 years. Other parts of the plant are also used by local Aboriginal people, including bark fibre for traps, nets and baskets and wood for weapon making. These uses are woven through local Songlines (Dreaming stories carrying knowledge and culture between and among generations), as are descriptions of deliberate carriage of the species by ancestral spirits (representing real people) from the east coast to the Western Ranges where they now also grow.

Although an analysis of linguistic evolution did not support or negate the hypothesis, reading aloud those local names recorded for this research conjure in my mind a soundscape perfect for contemplating the journey of this tree through the mountains: *junggurra, wirrum, yiwurra, baway, ganyjuu, mirrany, wanga, ganyjurr, mia, mai, boggum, bugam, binyjaalga, wiguuli …*

More prosaically, perhaps (although the nucleic acids coding for genes – *adenine, cytosine, guanine* and *thymine* – have their own resonance and rhythm), the DNA extracted from populations throughout this area show that the trees are the result of recent dispersal events, sourced from a small number of closely related plants.

The simplest explanation for this combined cultural and scientific evidence is recent dispersal by Aboriginal people. Alternative explanations for dispersal such as floods and other extreme weather events, or fruit-eating animals (today, or now extinct), don't stack up as well. They are possible, but less likely.

In a follow-up piece by Maurizio Rossetto and his colleagues,[75] they cite other examples of plants transported through Australia by its first human inhabitants. Despite the extensive destruction of evidence due to European colonisation, it is emerging that Australia's flora was not only dispersed by First Nations peoples but modified through their active selection of (cultural) traits.

This leads to the question of habitat restoration and what it is that we are restoring,[76] given the emerging extent of human interaction within nearly every ecosystem in Australia. For me, this same conundrum extends through to all human-assisted distribution and evolution, and it is something I consider in the next essay.

For this team of geneticists, botanists and cross-cultural ecologists, the regional origins of black bean at least are clear. They call for similar studies to test longer-distance transport of black bean from the north, something also inferred from other Aboriginal knowledge sources.

The gods, then, may have borne both palms, beans and much more.

April 2024

The weed-erness

There is no 'balance of nature', no 'natural order' and certainly no 'original sin in disturbing nature'

Many Indigenous Australians, and others, consider the concept of 'untouched wild areas' and 'wilderness' anachronistic, arguing that people have been and always will be part of nature, or Country. Humans have shaped the Australian landscape for at least 60,000 years, from cultural burning to the nurturing of plants such as the boab and Central Australian cabbage palm.

Overlaying, and most often disrupting, those intricate connections between First Australian people and (if you'll allow me) *First Australian plants*, are the catastrophic interventions of the last few centuries. What hope is there of unscrambling the colonial egg that is the Australian landscape as we find it today?

In her essay 'Seeing the wood for the trees', natural history author Danielle Clode takes a pragmatic approach to answering that question:[77]

> If only we could return to a pre-colonial ecology, free from the weeds, predators, and pests that have done so much damage to our environment. If only we could retract our own invaders back to their homeland. But

> that perfect world has never really existed, and we can only go forwards into a future we cannot predict, of unknown climatic variables, where northern pine trees populate southern forests, and Southern Blue Gums grow on every habitable continent on Earth. Sometimes we can fix restore, protect, made amends. Sometimes we just have to make the best of what we have in front of us.

This implies that we must, sometimes at least, accept that post-colonial plant immigrants are as much a part of the Australian flora as the First Australian plants.

In 2015, British journalist Fred Pearce published a spirited, and rather extreme, argument in favour of a world where (nearly all) weeds are welcome. *The New Wild: Why Invasive Species Will Be Nature's Salvation* demolishes any remaining pretence that the natural world can be restored to some kind of pre-human Eden.[78]

According to Pearce, humans have not only deliberately manipulated much of the world's vegetation – including most of the

Amazon forests – but human-assisted weeds are now integrated to such an extent into our 'wild' vegetation that their extraction would be very much akin to trying to unscramble an egg.

I have sympathy with this argument, although it perhaps applies differently in Australia, where we can identify extensive landscapes shaped primarily by pre-European forces – human and otherwise; places where few to no post-European settlement weeds thrive. While I too have questioned a weed-free perspective in previous essays, my concern has been around individual aberrations rather than the wholesale transformations documented and advocated for by Pearce.

Advocation is the appropriate word here because Pearce views this altered state as not only an inevitable world but a better one. Because plants from elsewhere – which Pearce calls *alien species* – increase the total number of species living in a particular habitat, the outcome, he says, must be a good thing. After all, that's what scientists and conservationists strive for when they restore damaged habitat: more species richness.

Well, yes and no. In the first place, adding extra species won't necessarily increase species diversity. Most of the severely 'weed-infested' (to use a loaded term Pearce would hate) habitats I know, become 'choked' (again, sorry Pearce!) with introduced species to the exclusion of native species. I'm sure this is variable from place to place, and that sometimes weeds don't displace more species than they add but it seems disingenuous to say that weeds always increase species diversity.

Then there is the supposed aspiration by all good folk for increased species diversity above all else. I would disagree with this and suspect so would many plant scientists. I'm sure many Brits are jealous of Australia's plant diversity, with their own depauperate flora a result of relatively recent Ice Age (20,000 years ago). As Pearce says, 'almost the entire flora and fauna of Britain has arrived in the past 10,000 years'. To me, a healthy ecosystem can have high or low species diversity.

The total number of species is not some kind of well-being index: a perfectly decent rainforest is rich in species, a desert poor.

That all said, I think Pearce is right to question the assumption that adding new species is always a bad thing. There are pests and weeds, aliens if you like, that do mess with our sense of *naturalness*. Pearce mentions the dingo in Australia as either an alien or something like an honorary native. The tumbleweed of the United States, *Salsola tragus*, is a Russian thistle that made its way to North America as a stowaway in flax seed, and the honeybee (responsible for 80 per cent of insect pollination in US agriculture) is a human-facilitated introduction in the Americas.

Conversely, he baits Australians by writing that the rabbit is 'almost a defining myth of the new nation' yet the 'real damage had been done by the farmers and their sheep'. Again, he has a point, although I wouldn't say that shouldering some of the blame means we should embrace the rabbit as an 'honorary native'.

Presumed native species can also warrant a little scrutiny before we declare them as all on the side of good and purity. Here in Australia, we are quite used to this, with species such as the Cootamundra wattle and sweet pittosporum weedy outside and even within their native range. Pearce mentions orangutans in Malaysia, saying many of them are 'probably descendants of animals kept captive for their meat centuries ago'.

Whether this is true or not – and I have no reason to doubt it – this points to a bigger issue around us already manipulating nature. How much and what kind of manipulation is allowed? Is that manipulation acceptable if it uses only indigenous ingredients, or perhaps if it seeks to restore an ecosystem to a particular composition, presumably defined by what we think it should or has looked like? You can see how things become morally problematic when we act as gardeners or zookeepers, but I'm not sure we can (or should) avoid this form of intervention.

Even if we can garden or tend a particular invasive species out of existence, this may not solve 'the problem'. Often weeds are symptoms of a broader ill such as nutrient enrichment, so simply removing them does not allow the 'wilderness' to return.

There may be boom-and-bust cycles, such as can occur with the marine algal weed *Caulerpa* in Australia, which in the end may benefit the 'natural system' by taking pollutants or excess nutrients out of the system. The science of these interactions and their longer-term impacts is generally lacking.

Then there is the matter of whether we should treat the human vector any differently to a seed disperser who happens to be a squirrel, mistletoe bird or ant. I've raised this in previous essays and I would simply add that there is now sufficient proof to argue that major plant groups, such as the protea and banksia family, Proteaceae, travelled across vast oceans unaided by *Homo sapiens*. These transits were long enough ago for the immigrants to be considered part of the natural flora of countries such as Australia.

The question then becomes is there a time frame, or species restriction, around what long-distance transport events are considered natural? Pearce demonstrates examples where plants and animals have travelled across areas as vast as the Pacific Ocean, arguing that therefore 'nowhere is too remote to be visited by aliens'.

Pearce's biggest beef is with what he sees as the ingrained irrational and unreasonable response many of us have to weeds and pests. If we could all dial down the anguish and fear, I think Pearce would, with some gentle persuasion, admit that weeds are not always good and may sometimes be a little evil.

He won't be drawn along that path, or tempted down from his soapbox, without acknowledging that native species don't always go extinct when weeds arrive, that ecosystems cannot (always, often, mostly) be 'fixed' by removing a weed, and that most invading species disappear relatively quickly (he quotes 10 per cent as hanging around to cause trouble). Pearce argues we have set up a false and unworkable dichotomy, and to place value judgements on species or communities – that is, good and bad – is a nonsense.

The reality is that we live in a world of compromised (again a loaded term from me) landscapes. Most, or all, our rainforests, says Pearce, are 'partly regrowth and partly a result of deliberate planting': they are, as he puts it, abandoned gardens rather than wilderness in the sense most people use this term. There is no 'balance of nature', no 'natural order' and certainly no 'original sin in disturbing nature', and weeds (alien plants and animals) are 'not, of themselves, bad things'.

Pearce can barely contain his rhetoric when it comes to comparing pristine habitats – 'increasingly like theme parks for conservation scientists' – with 'the unmanaged Badlands and novel ecosystems ... the bits of nature we don't cosset and pamper... the new wild'. He does acknowledge important questions, such as whether global biodiversity matters (he says 'to nature') and whether alien species destroy biodiversity through accelerating extinctions (or vice versa). As a salve to the bleeding hearts, Pearce accepts that 'nobody wants species to go extinct' and that he is hopeful the pace of extinction can be slowed.

To Pearce, species diversity matters a lot and alien species can help with this. His solution, contrary to what he sees as the current paradigm, is to support 'the new, rather than always spending time and money in a doomed attempt to preserve the old'. With this approach, proposals such as making the United Kingdom a sanctuary for species at risk of extinction – such as the introducing the Iberian lynx – are not fanciful. This, he says, is better than trying (ultimately unsuccessfully) to recreate some kind of landscape from the past and to try and hold nature in a false equilibrium.

In a small concession towards the end of the book, Pearce accepts that humans will want to intervene at times 'to preserve what we like and need for our own ends' and that 'we will sometimes need to defend against inconvenient invaders of our spaces, destroyers of our most cherished natural companions, pests and diseases'. And just in case you don't get it, 'we will continue to have our favourite species'.

How should we respond to Pearce's provocation? Putting aside whether Pearce's evidence and conclusions are sound in some or all situations, most hearts would tend towards the preservation of some kind of nature where humans have less sway or at least where they provide what we consider benevolent care. Then there are the competing demands for food and space, in a world where for many, poverty and health will overwhelm any bleeding heart.

The perspective is different from a place like the United Kingdom where the composition of vegetation is already an amalgam of plants that survived that last Ice Age, immigrants drifting in through whatever means since then, and extensive and wholescale deforestation.

In Australia, the landscape has been shaped by human occupation since well before the Northern Hemisphere Ice Ages, but those influences *appear* to have had less impact on species composition. Bluntly, until European settlement a few hundred years ago, most of the plants in Australia had evolved *in situ* for hundreds of thousands of years, even if for 60,000 years their relative abundance has been due to deliberate burning and forms of agriculture.

There are exceptions, as I've mentioned in earlier essays, and it is a matter of degree rather than kind, but I don't think a book like Pearce's would be written by an Australian. As to what sort of book would be written by someone living in the Middle East or India, I really don't know.

Climate change, of course, means that even if there was some inherent 'natural order' – and I agree with Pearce that this is a loaded and generally unhelpful term – that order is less of an idealised platonic state than it might have been. If we wish to retain the species alive on Earth today, deliberately moving species and gardening/zookeeping will be part of the new world order. We will be weeding and creating new weeds. Tending each collective of species, each ecosystem, to achieve some kind of predefined stasis.

Then there is the combined whole. Life on Earth. Most scientists and conservationists encourage connections between fragmented ecosystems to allow for change and interbreeding. This can also create a 'superhighway' for weeds, allowing invasive species to reach new frontiers more easily. But then Pearce may favour this as a way of hurrying up the process.

It does take us back to the objective of all this. Is it to increase the species diversity, no matter what? Is it to maintain the variety and rough proportion of organisms to some preconceived level? Is it to keep an ecosystem functioning, and if so, what does that mean – functioning as it does today or with higher productivity and/or diversity?

Sometimes there might be a relatively straight forward goal – like protecting a clutch of pretty, native orchids or keeping exotic pine trees from displacing native heathland. That requires us to understand the threats to the landscape and to be able to do something about it of lasting impact. I think Pearce would argue such situations are very rare, but I'm not so sure. At least in Australia. Yet even here, we can't always go back – the egg is often scrambled.

As Australian geographer Lesley Head put it,[79] 'profound thresholds have been crossed, and restoration of past environmental conditions is not possible'. She argues this doesn't mean we have to accept that weeds are part of every landscape (although like Pearce, Head would take issue with the view that any area can be weed-free, and whether that concept means anything much, given the way nature works), but that we must consider carefully where we can and should intervene.

Head compares the attitude and approach taken to weeds by home gardeners to how we view the wider landscape, honing in close to Fred Pearce's perspective. She concludes her essay on 'living in a weedy future' with the following, pertinent advice:

> ...we need a more open acknowledgement of the contradictions, edginess and difficult choices that attend contemporary Australian engagements with nature. The times require us to go beyond the ideal of a pristine past and more honestly face a fraught, unpredictable and surprising future. Resilient, opportunistic, larrikin weeds

> may be more useful companions on that journey that we can yet imagine – and gardeners who live with them one of our instructors.
>
> *Unpublished, June 2015*

Postscript

I drafted this essay in preparation for an interview with Fred Pearce on my ABC RN radio show (co-hosted with Jim Fogarty), *Talking Plants*.[80] In that live and lively conversation, we explored many of the ideas presented here and, as with his writing, Pearce was an engaging and persuasive guest. After reading *The New Wild*, I was clearly provoked, in a good way. The book made me re-examine my assumptions, which is always a good thing. Hopefully this response does the same for you.

What if weeds?

Alternative histories may make good fiction, but ...

I've never seen much value in the 'what if ... ?' approach to historical events. I can accept that alternative histories make good speculative fiction, but as a perspective on reality I find it dull and meaningless.

For example, you might ask – as many have – how the geography of Europe would look today if Napoleon had defeated Russia in 1812. But you might equally ask what the world would look like if, at 16 years old, Napoleon was accepted as a crew member on La Pérouse's ill-fated journey to Australia. And yes, he did apply.

Apart from not knowing which alternative tyrant or leader might have done or not done, I can (if I want to) equally imagine a history where the French ships *L'Astrolabe* and *La Bous sole* make it back to France rather than foundering off Vanikoro, thanks to our Corsican's brilliant leadership skills in a crisis. You can make up your own history.

When it is proffered that Australia may have become a French colony if La Pérouse had made it to Port Jackson or Botany Bay a few

days earlier, I shrug and point to the fact that the Spanish were building an armada of warships in Uruguay until that country's allegiance swapped to Britain so they could fight our unsuccessful La Pérouse crewman in Europe rather than storm the garrison at Port Jackson, or any other alternative scenario that might have developed, depending on a stray bullet here or successful cannon ball landing there.[81]

That all being said, I'm going to bravely (stupidly) break my own rule and propose my own 'what if ... ?' alternative history of Australia's weed invasion. The three often-cited incursions are: the misguided introductions by the acclimatisation societies of the 19th century; the propagation and display of exotic plants by botanic gardens; and adventitious entry of weeds along with introduced crop plants and livestock. There were and are many other points of entry, of course, but let me use these three as exemplars of the way an overseas plant of weedy potential might find its way into Australia.

First, those accursed acclimatisation societies. What were they thinking! Mostly, it seems, that Australia was a rather wanting landscape when it came to wild fruit and game. Lilly pillies and kangaroos didn't measure up against blackberries and rabbits. The whole ambience lacked a little something. And that something sounded like a blackbird, among other imported songbirds.

Even when happy with what nature had endowed to Australia, the acclimatisers saw that they could improve upon its placement. This meant moving wildlife into places where it would suit the new settlers better, such as transporting the Murray cod from its namesake river to the Yarra River catchment. As Eric Rolls puts it,[82] 'there was never a body of men so foolishly, so vigorously, and so disastrously wrong'.

The blackberry (various species and hybrids of *Rubus*) is a wonderful example of this misguided intent, led by the otherwise esteemed director of Melbourne's Botanic Gardens, Ferdinand von Mueller. In the 1850s, Mueller and others first championed the blackberry as the perfect plant to stabilise riverbanks but talk soon turned to its ability to produce the perfect bush food – sweet, juicy and tough enough to survive beside any byway.

Weed enthusiast John Dwyer quotes a lovely anecdote first reported in the *Victorian Naturalist*, recalling that wherever he roamed, Mueller would carry blackberry seed.[83] He said that after boiling his billy, he would often scatter a few seed near the ashes of his fire. His dream was to naturalise the blackberry 'on the rivulets of any ranges', knowing that then, 'poor people in time to come would bless him for his thoughtfulness'.

Still, as Dwyer also observes, the blackberry didn't need Mueller's help. It was planted in northern Tasmania and around Adelaide by the 1840s, in Sydney and the Bathurst region by the 1850s, and by the 1880s, was 'ruining agricultural and pastoral land' in the Illawarra where it had been widely planted as a hedge plant. With the aid of blackbirds (admittedly a connection back to Mueller), the blackberry was already off and running. There were no border controls or plant quarantine inspections into Australia, so the acclimatisation society simply hastened the inevitable.

With their brains and brawn and the riches of the gold rush, the Victorians wanted to achieve in a few decades what had taken the mother country centuries. While it was a daft idea, the settlers were schooled on how England had benefited from fruit and vegetable introductions by colonising forces from the Romans onwards.

Historian Linden Gillbank quotes Clements Markham – who, as she notes, led an expedition to steal *Cinchona* from Peru – proclaiming the introduction of useful plants from a distant land as 'one of the greatest benefits that civilisation has conferred on mankind'.[84] It increased material comfort and profit and was more durable than engineering, he extolled.

For Ferdinand von Mueller, plant acclimatisation was an essential part of the botanic gardens project. While he defined the primary purpose of a botanic garden as gathering the world's plants and then displaying them in a way that best demonstrates their various scientific and economic qualities, his words and actions betrayed a far more interventionist role in shaping a country's landscape.

Mueller wrote extensively on the need to conserve native vegetation and the diversity of plants contained within, but also on the benefits of adding to that natural bounty. During his time as Botanic Garden director and Government Botanist, he distributed thousands of plants and seed to botanic gardens and individuals throughout the State. Mueller was, after all, a founding member and, for 12 years, vice-president of the Victorian Acclimatisation Society.[85]

Two-thirds of the plants now naturalised in Australia have escaped from gardens, including 14 of the 20 Weeds of National Significance in 2006, and more than a third of Australia's 'declared' weeds are invasive garden plants.[86,87] Botanic gardens have clearly been part of the process, actively introducing into Australia thousands of plants from overseas. In some cases, their directors have championed the planting of what were to be come environmental weeds. Not only Mueller: in the 1920s, the curator of the Botanic Gardens' Herbarium in Sydney recommended the Cootamundra wattle (*Acacia baileyana*) as a plant perfectly suited to the home garden.[88]

Yet while botanic gardens attract a lot of the flack, most of the actual or potential invasive plants are available to purchase in domestic plant nurseries.[89] This includes species declared as noxious weeds or recommended for eradication nationally. Their presence in botanic gardens would have marginal impact, apart from the lost opportunity to educate and inform the community about what plants they should and shouldn't grow.

That, to some extent, is being done and botanic gardens today would not display (without some careful management) or encourage growing a potential agricultural or environmental weed in their region. Botanic gardens also review all new acquisitions for their potential to become weedy, with tools now available to assist with this process.[90] Still, it is fair to say that without botanic gardens, particularly in their early years, weed introductions would have been slower. We may well be at the same point today but that is a difficult 'what if' to assess.

While plants deliberately introduced into gardens are clearly the major source of weeds in Australia,[91] there are other easy routes into the country. These include as castaways with imported stock and other wildlife, contaminants in a range of plant and animal products (often for food and medicine), attachments to our clothing, and various non-human mechanisms such as drifting in by seas and floating through the air.

Australia has some of the world's strictest quarantine procedures in place to control these entry points, but really, it is more a matter of 'when' rather than 'if'. Still, the longer you keep a plant or a pest out, the more chance there is of finding or putting biological or other controls in place to further delay or restrict the extent of impact.

Dutch elm disease is a good example. We have the vector, the elm leaf beetle, well established in our imported elm trees, so when the fungus (species of *Ophiostoma*) arrives one day in a plank of oak wood – as it will – the disease will take off. So far, the fungus has been effectively kept out of Australia, and by the time it arrives there may be a treatment available and certainly more resistant elm variants to replace those most at risk. This is good to know given the 11,000 or so English and Dutch elms in the inner suburban streets and parks of Melbourne are now among the largest stands of mature elms in the world.

Despite all the historical records in the National Herbarium of Victoria of plants established near Coode Island Quarantine Station, with tighter quarantine controls introductions through agriculture are less likely today. As with acclimatisation societies and botanic gardens, unregulated importing of agricultural plants and animals in the colony's early years will have accelerated the rate of introduction of environmental and other weeds in Australia rather than increased the total number.

As mentioned earlier, that is still problematic and should have been avoided, but like the consequence of the French or Spanish – rather than the English – colonising Australia, who's to say things would be better today.

April 2024

3

Garden plants and landscapes

Can you maintain and honour a landscape design after the inevitable demise of the designer and the equally inevitable decline and death of its constituent plants? At one extreme, the design might be viewed as colour-by-numbers outline, where future garden carers and owners simply recolour it now and then, as per the instructions. Like for like, in terms of species composition and placement. Curve for curve, and vista for vista, in terms of outline. Nothing wanted or needed in terms of creative input.

This assumes such an approach is possible. Plants grow larger, change in shape and form, and then die. Seldom can you replace some landscape element with a fully sized replacement, nor should you, for the stability and health of the new plant. Overlay that with changes in climate and surrounding environment which may alter what will grow successfully in the garden.

Then there is garden fashion or personal taste. Should that have any place in a heritage garden? I think yes, which brings us to the other extreme. Anything goes.

In this scenario, the original designer is treated as muse perhaps, or simply setting the tone and style but not the detail. This is more analogous to an art movement approach. Our garden is Art Nouveau, Post-Impressionistic, Old Master and so on; or if you prefer, baroque, picturesque or gardenesque. Abide by a few rules then do what you like.

A happy middle might be struck by adhering as closely as possible to the design intent and aspiration. For a home garden, this might be a sense of geometry and relationship to nearby buildings or other landscapes. For a botanic garden, its role as a place of beauty as well as science, learning and conservation comes into play. For a grand garden or park, something more akin to the botanic garden philosophy, we might put ourselves in the shoes of the designer but on today's turf. In all cases, we could engage a 'like-minded' designer to guide us.

To torture that art analogy a little more, gardens are not generally, I think, like a finished work of art, to be restored and preserved as close as possible to the original. They are more like a piece of music or theatre, to be reinterpreted and reimagined as desired. Sometimes, of course, they may be kept (performed) in their original form. If you want. Eventually, like a building, the garden must be propped up and repaired, or replaced. Never, I would suggest, reconstructed in faux style unless that is the eccentricity you want to feature.

British author and historian of gardens (and confectioner), Tim Richardson, weighed into this debate in 2014, with a typically refreshing and challenging essay for Australian audiences.[1] For 'highly personal' historic gardens, he favours the 'like-minded' designer approach, or indeed an 'equally talented' designer. For a garden of more generic design, Richardson's preference is repair over restoration.

I mentioned Richardson's perspective in a blog post on Roberto Burle Marx, posted after a visit to his garden (*sitio*) on the outskirts of Rio de Janeiro, Brazil.[2] The current owners of that property were faced with a decision about repairing what was there or pursuing the spirit of the original garden as a creative and experimental space. When I was there in 2015, they were flirting with inviting new designers to pursue Burle Marx's intent.

Australian garden writer Kim Woods Rabbage commented on my blog at the time with a note of caution. For a designer of such importance, she said, some of his original work must be kept intact. This is in the same way we want to see the work of artists like Picasso, Matisse and Warhol, rather than (or in addition to) those they have inspired. In my reply to Kim, I said I suspected most of the garden (90 per cent was my estimate) would remain in his original design but they should have the confidence to add 'a bit here and there with the same sense of adventure and creativity'. I don't think you can ever polish and clean a garden as you would a stone monument.

The plant palette used to restore or repair an historical garden can also be contentious, and with climate change that choice becomes even more vexed. Although Burle Marx was keen on encouraging local plants into gardens around Rio de Janeiro, he was happy to use beautiful plants from anywhere if they suited his design. I wrote in Chapter 2 about why weeds might be encouraged or at least less scorned, but it is also worth testing those biases we might have for plants outside our region or country.

George Seddon, in *The Old Country* (2005), reminds us that kangaroo grass (*Themeda*) grows naturally in South Africa *and* Australia. The tropical Australian boab (*Adansonia*) has its closest relatives in Madagascar and southern Africa.[3] And most Australian conifers and podocarps in Tasmania have their closest relatives in New Zealand and Chile.

Indeed, of the 19 species of *Araucaria* – an Australian conifer – three are native to Australia and 13 from New Caledonia. From the same plant family, our single Kauri (*Agathis*) in Australia is matched by one in New Zealand and several in New Caledonia. These international relationships, said Seddon, should be celebrated rather than scorned. Yes, we have a fascinating and distinctive flora but why not be inclusive rather than exclusive. As I'll argue in one of my essays,[4] if you live in Melbourne, a plant from New Caledonia is closer to you than one from Western Australia, and as Seddon observes, may well have as strong or a stronger botanical connection to your local flora.

In my recent memoir,[5] I landed on charging the care and repair of a heritage garden to 'good people of sound mind, with the right expertise, making a judgement call on when and how to intercede'. I'm happy with that still, but here are a few things they might consider.[6]

Guiltfree planting

Plant not a new weed, nor one in too much need

If, as I argued in the last chapter, weeds are here to stay, why not learn to love them? We could even learn to garden with them. Many of us accept the occasional weedy plant erupting spontaneously in our garden path or bed, if it's pretty enough.

Personally, I don't mind sweet alyssum (*Lobularia maritima*) and kidney weed (*Dichondra repens*), and I remember enjoying a pink jasmine (*Jasminum polyanthum*) that weaved its way over and through our neighbour's fence. Perhaps the odd cheery dandelion, or a blotch of yellow or pink shamrock (*Oxalis*), in my lawn?

Landscape designers in New York have taken this a step further,[7] adding weeds to their planting palette. In 'a garbage-strewn industrial lot covered in weeds' in Ridgewood, Queens, not only were the new owners 'inspired' by the spontaneous plants (weeds), they propagated and replanted more. Designing with weeds, as they called it, created a pleasing landscape *and* reduced the effort required to establish and then care for the garden.

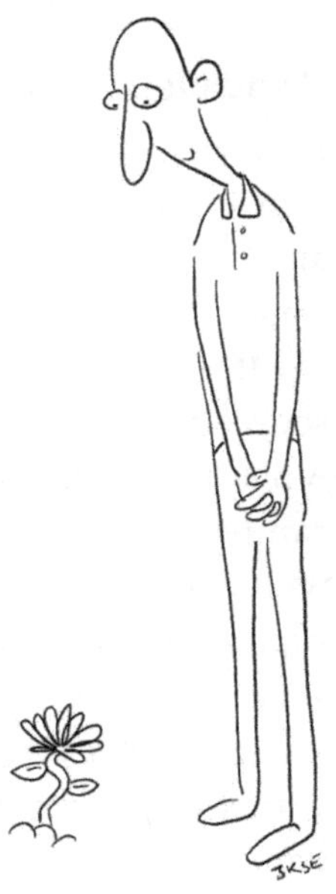

This all makes sense, although there are weeds and there are 'weeds'. In their case, the New Yorkers inserted some new introductions among what turned up spontaneously, and among the spontaneous, some weeds were more equal than others. 'Native weeds' – species apparently indigenous to the area before human settlement but still finding purchase in the urban streets of New York – were favoured.

Which got me thinking about what advice I might give to someone who wanted to plant out their home garden in an environmentally responsible way. Not so much which species but what to consider when selecting (or encouraging) plants that do 'no harm'. I came up with four principles.

Principle 1: Not a new weed

Generally, it doesn't matter if you plant what might be called 'old weeds', species that are well established in your area or come in unaided anyway. There is some nuance here clearly, and if there are efforts (or requirements) locally to remove and discourage a particular species, you should (or must) follow suit.

What you don't want to do is create the opportunity for a new weed to become problematic outside your garden – whether that is in other people's garden, cultivated land or nearby bushland. Find out as much as you can about the species you'd like to add to your garden, then decide whether you are willing to take, and manage, the risk.[8]

Principle 2: Nor a plant in too much need

Be wary of any plant that requires more than your climate and soil can provide unaided. In most parts of Australia, that means avoiding anything that, once established, requires the addition of 'too much' water. With some caveats.

If you have a rainwater tank, recycled water system or sustainable source of ground water, you can be more generous with your watering. A naturally damp area in your garden, perhaps an old drainage line, can be a good place to locate water-loving plants. You may even be willing to cart buckets of greywater from your shower or laundry to support a few favourite plants. Generally, though, favour plants that survive on natural rainfall or seepage.

And water is just one requisite for a healthy garden. It should go without saying but avoid planting anything that needs environmentally damaging chemicals to survive or thrive. Plant nothing that will fail without the addition of toxic pesticides or excessive fertilisers, which means avoiding species susceptible to common pests and diseases in your area.

That said, nibbled leaves and a few tatty stems are usually the symptom of a healthy and biodiverse ecosystem. Tolerate, and even celebrate, a little wear and tear in your plants.

Principle 3: Location, location, location

This principle could also be expressed as *look around you, and don't do anything stupid*. Where you plant something – its location – matters on many scales. First, the soil and conditions where you want the (usually) small plant to establish – you can generally manipulate this to some extent. Secondly, the envelope likely to be occupied by the mature plant – again, you can allow for this, but you must consider it at the time of planting. And thirdly, the climate and environment at your place of planting – something you normally can't change but must consider.

Considering these three elements of position is particularly important for long-lived plants such as trees. For example, is there room for its canopy and its roots at full size? That means obvious things like not planting tall trees under power lines or locating invasive-rooted species next to your house or plumbing.

Consider the climate likely in 50, 100 or however many years your plant may live. With climate change, we expect Melbourne to be dryer and hotter than today, with nearly a fortnight more days every year over 35°C by 2070. Will your long-lived tree survive and thrive under those conditions? For all plants, will they attract or detract local wildlife? Which of those you want is up to you.

While local plants are likely to be good for attracting local wildlife, exotic plants often do as well. It is usually more important to have a mix of plant forms, including grasses and ground covers, dense shrubs, and trees. In fact, planting very local species has its own problems,[9] but in the context of massive habitat loss, invasive plants and animals, and climate change, its generally a good thing.

Principle 4: Neither plane nor plain

Finally, I hate to jump on the 'I hate plane trees' bandwagon but my issue is not with them shedding fruit hairs (noting their pollen is unlikely to be a problem) but with their dullness. Not dull as an individual tree; there are some spectacularly grand specimens in the streets and older establishments of central Melbourne. My gripe is their ubiquitousness.

With good reason, planes are a commonly planted street tree. They are long-lived and shady, and they tolerate pollution, poor and compacted soil, as well as poor pruning. Plane trees might be described as an engineering solution to street tree planting – effective and efficient but, no offence to engineers, a little dull.

My recommendation is diversity in planting. That will not only create an interesting and beautiful landscape – yes, you can and probably should plant more than one of each species – but one that is more resilient to pests, diseases, and climate change. If (when) Dutch elm disease, ash dieback or cinnamon fungus arrive, you won't lose all your trees. You will also be able to enjoy some variation in leaf form, flower and seasonal shadiness.

If you want to know what grows well in your area, look at local parks and gardens, particularly the neglected ones where species survive with less mollycoddling. On the other hand, some experimentation can be fun.

The second director of Melbourne's Botanic Gardens, arguably the greatest botanical garden landscaper of all time, William Guilfoyle, once said:[10]

> It should ... be the aim of those in charge of public gardens not to reproduce vegetation which may be seen in other portions of such gardens, but to bring before the public, in special spots, scenes of beauty not to be found elsewhere, by representing plants of a different character to those more or less common to the locality.

You don't have to do that in your home garden, but you might.

Unpublished radio notes, May 2003[11]

Postscript

A question that often comes up when I speak on this topic is whether there is enough variety in 'the trade'. With more than 36,000 plant species in cultivation in Australia[12] – some 11,000 native to Australia and the rest introduced from overseas – there should be plenty to choose from. These calculations come from weed guru, Rob Randall, a not dispassionate onlooker, adding this was 'an amazing number of plants available to horticulturists, landscape designers and plant lovers in general'. That is, in Randall's view, enough and we don't need more.

Coincidently, there are as many non-Australian plant species in Australian gardens – about 25,000 – as species growing naturally in the entirety of Australia. While 10 per cent or so of the garden species have become naturalised[13] in Australia – adding nearly 3,000 to the total number of 'wild' species in Australia – Randall says another nearly 6,000 have 'form' overseas as weeds and should be avoided in areas where they might reasonably be expected to grow unaided in your climatic zone. That still leaves us some 16,000 exotic plants already here in Australia with presumed low weed potential.

If we add back in those Australian species Randall lists as in cultivation already (noting that some of these will have their own weed potential in parts of Australia), this gives us a palette of more than 27,000 species for our home garden. And that is not including subspecies, varieties, forms and cultivars.

Some species will be suited only to parts of Australia but I would guess that at least half of them, with some mollycoddling, will grow where you are (unless you are in one of the more extreme climatic zones in Australia). For arguments sake, let's say as a Melbournian there are at least 14,000 plant species available to you.

That might seem like plenty to choose from if only nurseries propagated and stocked them. Given the environmental damage caused by those that do escape, should we not be content to potter with what we have? Well, yes, but you could argue that more would be better.

First, collecting and seeking the new (or the full 'set') is an obsessive human pastime and not intrinsically a bad thing. Connected to this, there may be beautiful (awe-inspiring?), interesting or quirky plants out there that if growing in our garden would cause us to smile or think a little. Again, not a bad thing.

Secondly, there may be plants better suited to our gardens, requiring less intervention on the environment to keep them alive perhaps. These may be individuals or variants of a species already in cultivation, or a new species entirely. With accelerated climate change we might want to keep plant species or forms that we like but select variants, or replacements, better adapted to our new growing conditions. James Hitchmough, from Sheffield University, has preached this line in Britain, travelling regularly to China to select new hardy plants for UK gardens, plants that will grow better now or when the climate changes further in the future.

If we stick to my four planting principles and are circumspect and vigilant when introducing a new plant into cultivation then I think we should continue seeking novelties for the garden, for the same reason we write and read new books, seek new artwork and travel to places unseen before.

Similarly, we should value and care for what we have, safeguarding those species already in cultivation (including all their variation and cultivars). Let's not reduce, through neglect or desire to follow a fashion, the wonderful variety of plants we can grow in Australian gardens.[14]

Immortal plants don't live forever

> *A tree can be viewed as a close-knit colony of many individuals, a colony with the potential to never die*

I was shocked and awed by the provocations in French tropical botanist Francis Hallé's wonderfully titled book, *In Praise of Plants*.[15] I was also amused and captivated – it's a fascinating book. Hallé is particularly

keen to show that most plants are fundamentally different from animals, and that we can't simply generalise what we know about animal biology to plants. To my delight, Hallé also concludes that plants are far more interesting!

One of Hallé's key concepts is that a tree can be viewed as a close-knit colony of many individuals rather than a single organism, and that this colony has the potential to live forever. What he means is that there is a repeated pattern, and each unit can continue to grow (whether part of the tree or as a cutting or graft) as long as it contains a bud.

The bud, according to this interpretation, can be considered the true individual – it cannot be divided any further. Even if we use genetic uniformity to define an individual, in a long-lived tree a slow build-up of mutations in vegetative cells can result in some branches having a distinct fingerprint. So, a tree is like an ant nest or sea anemone: individuals die but the colony persists.

Apart from providing an interesting linguistic or philosophical exercise, does any of this matter? It does if you consider colonial organisms to be, to all intents and purposes, immortal. Clearly most trees are mortal: Australian wattles tend to flourish and die within a decade or two, and even our most majestic street trees have a maximum life span of one or two centuries. Structural problems develop. Food and water supplies can't be guaranteed. Fungal pathogens somewhat shortsightedly kill their host. Wind knocks them down. And so on.

Perhaps this is why the longest-lived plants are not the tallest. A nearly 5,000-year-old bristlecone pine (*Pinus longaeva*) in California, felled by a park ranger in 1964, is often cited as the oldest tree. Like all bristlecone pines, it was less than 6 metres tall. There are claims that others of this species, still alive, may have begun life 8,000 years or more ago.

A sprawling Huon pine (*Lagarostrobos franklinii*) near Mt Read in Tasmania extends over 2.5 hectares (the size of a city block), but the individual trunks are typical for the species at 10 to 20 metres tall. This massive plant is estimated to be about 10,000 years old; older than any of those bristlecone pines, although perhaps of a similar age to Norway spruce (*Picea abies*) growing in Sweden. Here's the rub though. This Huon pine, and those Norway spruces, are 'colonies of clones', not a single tree like the bristlecone pine.

If we accept Hallé's view that most plants are colonies anyway, we shouldn't care too much if trees like the Huon pine survive only because they spread vegetatively at their base by producing new stems to replace old (i.e. they sucker or layer). This also brings into contention a creosote bush (*Larrea tridentata*) from California, now over 11,700 years old. But this is a baby compared to a strange plant lurking in the Tasmanian World Heritage Area.

Some years ago, I took part in an expedition sponsored by the *Australian Geographic* to Bathurst Harbour in south-western Tasmania. While I was wading through tea-coloured streams searching for new species of red algae (as I do), Jayne Balmer from Tasmanian Department of Primary Industries, Water and Environment was collecting samples from one of Australia's oddest, perhaps its oldest, and certainly one of its rarest, plants.

A member of the Proteaceae family and closely related to waratahs (*Telopea*), King's holly (*Lomatia tasmanica*) was first noticed by European settlers in 1934, when local identity Denny King found it not far from his home in Melaleuca. Those plants are now assumed dead, but King found a second 'population', confirmed by Tasmanian botanist Winifred Curtis in 1965.

This new cohort looks healthy enough, extending along creek gullies for over a kilometre, but none of the plants produces fruit or seed. Genetic testing by Jasmyn Lynch (who works with Balmer) and colleagues from the University of Tasmania showed no detectable variation across the entire population.[16] This is usually good evidence of a vegetatively reproducing species (although some plants that grow from seed may be genetically indistinguishable from one another, such as the Wollemi pine, and, like the branches on an old tree, vegetative off-shoots are not necessarily genetically identical).

Microscopic examination also showed that King's holly has three sets of chromosomes. When it comes to chromosome numbers, plants do mix it up a lot, and multiple copies are not uncommon. But triploids, as they are called, are rare. In *Lomatia*, and in fact in all its close relatives, a double (diploid) set of 11 chromosomes is standard issue.

The fact that King's holly has 33 chromosomes explains why it can't produce fertile seed – triploid plants rarely find a way to split this odd number up and produce viable gametes (the reproductive cells that have a single or haploid set of chromosomes). This odd set of chromosomes probably resulted from the successful fertilisation many years ago of a freak diploid gamete with a normal haploid gamete. Two plus one equals three!

Lynch and colleagues hypothesise that every plant in the 1-kilometre stretch was once connected, and that fire has probably fragmented the 'clone'. Based on a combination of its current extent, carbon-dated fossils, the lack of genetic diversity, absence of seed and the unlikelihood of triploids occurring twice, they estimate the clone started life over 43,000 years ago. Hard to confirm but a tantalising proposition.

But wait, there is more. A few thousand kilometres north, plant ecologist Rob Kooyman found another long-lived clone. He suggests, provocatively, that the peach myrtle (*Uromyrtus australis*) in New South Wales's Nightcap Range is, at least functionally, an immortal plant. That is, in the right circumstances it could live forever. Kooyman and his research supervisor, Peter Clarke from University of New England, are still trying to confirm the exact age and life history of this intriguing plant.

Like King's holly, the peach myrtle has found a way to survive without reproducing from seed. While its small white flowers develop into black, berry-like fruits with plenty of seed, there is no regeneration from seed in the wild. Interestingly, Kooyman can get them to germinate easily in nursery pots.

In the Nightcap Range, each individual consists of a large group of stems up to 12 metres high, the biggest of which seem to be about 1,500 years old. The plant 'regenerates' by replacing old stems with new and is likely to be at least 10,000 years old. Not quite as old as King's holly but if Kooyman is right, and destructive fires are kept out of this area (see Postscript), it will get to any age you care to nominate.

This does , however, exclude the real stayers of the non-animal living world. Giant fungal networks are said to be the largest living organisms on the planet, and possibly functionally immortal as well. Plenty of algae, fungi, bacteria and other microbes reproduce almost exclusively by splitting in two (without any sexual fusion), and you could describe their extended families as exceedingly old but disjointed individuals.

This casts a dark shadow over the paltry efforts of most animals, which at best live for a few hundred years or, if you are a sea anemone, a couple of thousand years. Being a plant, or a microbe, has its benefits.

Nature Australia, January 2006[17]

Postscript

The potential immortality of plants is a topic I've written and spoken about many times, culminating in this essay in 2006. Almost every time I raise the topic, there are new contenders suggested for the title of world's oldest plant. Recently my friend and colleague Neville Walsh suggested that 10,000 years would be a reasonable estimate of age for the 'fairy ring' mallee eucalypts in southern Australia, which often lack their original central stems.

Self-proclaimed plant messiah (but also kind-hearted friend and supplier of indoor plants for our home when I lived in a house within Royal Botanic Gardens Kew), Carlos Magdalena, says *Tillandsia* growing in the dry sands of Sechura Desert in Peru are arguably 14,000 years old. I'm not sure if you'd class these drought-tolerant bromeliads as colonies or collectives.

More provocatively, it was pointed out to me that most commercial bananas are only propagated vegetatively, so perhaps we could consider all of them to be part of one very old individual. Perhaps. But that doesn't feel right. We could stipulate – for sake of argument – that an individual plant has to be physically connected for its entire life, or at least be in the place where it once had that physical connection. That means the oldest plant on earth (keeping in mind fungi are not plants) may well be King's holly in Tasmania – until, that is, we find another older collective of stems with a shared and intimate history.

Which brings us back to Rob Kooyman's peach myrtle in the Nightcap Range. Kooyman's colleague and highly respected genetic conservationist at the Australian Institute of Botanical Science, Maurizio Rosetto, posted the following on social media on 15 May 2021: 'Over a year since the Nightcap fires, impacted rainforests are struggling. Rare trees (*Uromyrtus australis*) are still dying and if basal shoots are present, they are often infected by myrtle rust.' Until recently, fire was considered a 'rare and unlikely' event in high altitude, wet forest of the kind where peach myrtle grows. The 2019–20 bushfires burnt 5 per cent of the known range of 800 to 1,000 'individuals' of peach myrtle and, as Rosetto said, much of the resprouting is now threatened by myrtle rust.[18]

If peach myrtle survives fire and pestilence, just how old are the oldest stands? When I wrote the original essay, Kooyman estimated individual stems might be up to 1,000 years old but he needed more information on growth rates and genetic population studies to estimate the age of the entire potentially colonial collective. I'm not aware of any further results yet, and in some ways I prefer the uncertainty. If we knew, I'm sure someone would tell me that there is another plant just a little older, depending on your definitions.

Tree branches ahead

Another street tree bites the dust

During the mass intake of breadth that accompanied the news of Donald Trump reaching the White House in 2016, a tree fell. Not in a forest but in a suburban street, and I definitely heard it.

Early one evening, an articulated container truck manoeuvred, badly, through our narrow but leafy Melbourne suburban street, clipping a 50-year-old pin oak on the opposite nature strip. That clip brought down half the tree – a few tonnes of wood I expect – over the truck's articulation point and the front of my car.

Looking out our front door a few minutes later, it was difficult to make sense of the odd protrusions of green in places where they shouldn't be. But first things first. Was anyone hurt? Apart from the

pride of the truck driver and potentially me, as I weighed up the inconvenience and cost of repairing my car, no.

Property damage? As it turns out, the end of the very large branch that made its way through the truck articulation to my car rested on my windscreen. 'That's lucky', said the policeman who turned up presently, 'it's the toughest part of the car'. I guess that's right because the branch didn't penetrate or even scratch the glass, avoiding costly repairs to those flimsy bits of metal that make up the rest of the vehicle.

After the State Emergency Services (SES) arrived and chainsawed through the oak limbs 'resting' on my windscreen, I backed the car down the street and discovered not even a scratch on the Duco! The SES then set to work on the rest of the tree, eventually allowing the truck to unblock our street.

The debris was piled up neatly for removal the next morning. As always, a tree, or half a tree in this case, seems to take up the space of a small forest when reduced to moveable-sized segments. The Council, I presume, came along later in the day and removed the rest of the tree, leaving a yellow-painted stump as a reminder of the once grand sentinel across the road. So, no damage to people or 'property' but one hell of an impact on this individual tree.

How do you value that loss? You can estimate the economic benefit of a tree's contribution to stormwater runoff, air quality, energy use and airborne carbon, which in the case of a street tree in Adelaide is at least $171 a year.[19] This seems rather low, and a better measure might be an additional $50,000 or so added to your property value when the leaf canopy of street trees is increased by only 10 per cent.[20]

You can also value the amenity of the tree. Royal Botanic Gardens Melbourne recently valued its 2,800 mature trees at $56 million – an average of $20,000 per tree – at the same time suggesting that an alternative method considering the 'splendour' of some specimens might double this total.[21] I am assuming that splendour would take more account of heritage and landscape significance, not just market value of the species inflated by its current bulk, age, form, vigour and suitability of location.

It is easy enough through these methods to arrive at rather large – and perfectly reasonable – numbers for each tree, but that's not what concerns me here. How do you value, in an ethical or moral sense, or perhaps more correctly how do you deal with, the loss of a 50-year-old tree?

You can't replant a 50-year-old tree. Trees in various stages of maturity can be selected to take its place, and in theory you can transplant a fully mature tree. But it is almost always better to select the smallest tree possible if you want it to be long-lasting and robust.

No-one wants human injury and at the time I was in slow recovery from minor surgery. So I was sensitive to pain and discomfort and wouldn't wish it upon anyone. Yet I also knew that I, also 50 or so years old, would probably recover. Every bump and jolt knocks us back a bit, but it's not like a 50-year-old tree returning as an entirely different 5-year-old sapling. Death and permanent disability, of course, are entirely different.

I would have been furious to have had to repair my car, having just done it recently after a minor scrape with a car-parking pillar. Cost, inconvenience and lots of paperwork. But the car could be repaired, and quite probably brought back to the standard of a typical car of its age.

A mature tree is another thing altogether. While there may be some financial penalty due to the Council by the truck company – I hope so because that seems entirely reasonable and expected – there is no way any amount of money will bring back a tree. Just as when a vandal ringbarks a tree on the coast to protect or create their view, this action cannot be undone (and covering the remains of our tree with black cloth or erecting a billboard is only a partial solution, or punishment).

Still, let's be careful not to see this entirely through my botanically rose-coloured glasses. This particular oak was not in great health. The break showed a structural weakness and the years of pruning to misshape it around power lines had done it no favours. It may have only had a few years or a few decades left before an arborist deemed its time had come anyway.

That's fine, but surely a planned death is a very different thing. A considered tree removal and replacement, as must happen from time to time. In this case, human error, unnecessary risk taking or perhaps a total lack of care for our roadside vegetation, led to an unfixable consequence.

I'm not particularly interested in blame but I do think instilling a greater appreciation and value on each of our precious street trees might make such tragedies less likely.

HMAA News, June 2017[22]

Postscript

Trees in public places have more to fear than oversized trucks. In the last week, as I write, there have been media reports of newly planted trees in a Victorian country town being snapped or ripped from the soil by wayward lads. In Sydney, yet more mature trees were chainsawed or poisoned to create unrestricted harbour views for local residents. In Adelaide, developers lopped nearby street trees as they prepare to fill each property with bricks and mortar.

It's a tough life for a tree living on the streets. In my neighbourhood, there is a sturdy bluestone retaining wall constructed around one camphor laurel, packed full of agapanthus. Now I dislike agapanthus at the best of times but here it is the soil built up around the tree trunk that seems problematic. I am all for nature strip gardening – and I do it myself, with some indigenous grasses and herbs – but let's also care for the trees.

Then there is the pruning regime needed to keep the paths and roads free of obstruction, utility wires and poles clear from the canopy, cars and people safe from falling limbs and, in the end, some kind of streetscape aesthetic. The typical pin oak topiary in my street resembles a weightlifter doing press-ups with two halves of the canopy. Rather odd and ugly but still better than no tree. Thankfully some councils have discovered smaller trees and shrubs, plants that can exist almost entirely below the powerlines.

In the end we want councils and communities to value their street trees, doing what they can to protect and nurture them. When one falls or is damaged, let's repair or replace it swiftly. I'm happy to report that two new oaks have been planted on the nature strip near to where the 50-year-old tree met its demise. Both are healthy and vigorous – almost reaching the powerlines.

Killing a tree makes room for more life

It is hard to please everyone - or anyone - when managing city parks and gardens

Kerry Packer dumped a truckload of mulch in protest. Sydney's Lord Mayor demanded I be sacked. Broadcaster Alan Jones described me as a fat, lazy bureaucrat. As I reflected in my memoir,[23] I wasn't too fussed

by the mulch (which I put to good use) or for that matter the opinion of the Lord Mayor, but I was mildly offended by being called a bureaucrat.

That was 2004, when I had the turpitude to remove 10 trees from Sydney's Domain, under my care as Director of the Royal Botanic Gardens in Sydney. I had only been Director for a few months and the tussle to remove the trees became a test of my resolve and resilience, as much as anything else. More importantly though, it was a test of how we, as a (mostly) relatively recent immigrant community in Australia, cope with our first large-scale replacement of elderly trees in towns and cities.

I was recalling the events a decade later as I prepared for a talk to an *Australian Garden History Society* forum at the State Library of Victoria called 'Trees – Natural and Cultural Values'.[24] Have things changed, I wondered? Can we now approach the demise of a well-loved tree with respect and dignity, accepting that our generation may have to live without its shade if we are to provide for the next?

Yes, I announced at the forum, Australians have matured in their thinking, partly because of the fuss made around the removal of those Moreton Bay figs in Sydney's Domain 10 years earlier. That 'debate' allowed for ample airing of the social, scientific and safety arguments for and against removing ageing trees. Most people now understand that while a tree should be cherished, it is not sacred under all circumstances.

The Burra Charter for conserving heritage places (which includes gardens and trees as well as buildings) is now cited regularly in both cities. On the downside, many civic leaders still value popular opinion more than expert advice. In my talk I added that Melbournians had always had a more mature approach to tree decisions, as it has on most matters (I had by then left Sydney for Melbourne, via 2 years in London, and was clearly joining in on the long-standing rivalry between Australia's two largest cities).

When I rang Kerry Packer to thank him for the delivery of mulch, he told me of magnificent old fig trees in Buenos Aires which I had now seen and admired myself. We prop and fence a few old favourites in Australia, as they do overseas sometimes, but that's only part of the solution. Mr Packer said if he was in my position, he would reconsider the decision to remove the Moreton Bay figs in the Domain. It was a cordial and mutually respectful conversation, perhaps best captured by his signing-off remark that I should remember his horticultural advice was worth what I was paying for it.

Alan Jones and I didn't speak at the time, although I was keen to go on air to say why I thought it was the right thing to remove and replace these particular trees, at this particular time. I would have explained how the trees were in poor shape, suffering from historical neglect and regular insect (psyllid) attack and that they were considered unsafe or soon to be so. Fencing them off wasn't possible due to their proximity to roads, their place in the landscape (as shade trees and part of the public parkland) and the unattractiveness of a Domain criss-crossed with fencing.

Apart from all that (and I would have been off-air talking to myself by then), there was the ethical matter of the trees having to be removed at some point and it being the responsible and generous thing to do now. Oh, and the need to remove an avenue in its entirety so that you can replace with even-aged trees (experience elsewhere in Sydney showed that you can't plant new figs near old and expect them to grow successfully, even where there is room for a second row).

The Lord Mayor of Sydney had her own tree problems soon after this, with a report to Council showing that much of the canopy of Hyde Park would need replacing over the next decade or two. Like Melbourne, Sydney has lots of parks and streetscapes made beautiful by old trees. Some of the most beautiful and beguiling were then a century or more old. Deciding when and how to remove and replace them is (to use that misused but in this case pertinent term) *core business* for councils and botanic garden directors.

I don't want to trivialise, or anthropomorphise, a topic imbued with strong emotion by equating the passing of a tree with that of a dying friend but some of the same dignity of life arguments can be made. It is not always best to maintain life at all costs. Death is unappealing, for us and for the trees we enjoy but keeping them alive isn't the only consideration.

Old trees can become unsafe, they can become unattractive in the landscape – not, I rush to add, unattractive themselves in old age (as with people, they can become expressive and if they age well, more attractive in maturity) – and in a city they have a finite life.

Like much in life, and death, it is a nuanced thing. As I said often 10 years earlier, a tree in a park or garden is not growing in an old growth forest. It is hardly ever the aim of the person planting a tree that it decays into habitat and regenerates in some way. We need park managers to make informed decisions in the best interests of the community now, and a few decades hence.

Saying it is right and responsible to remove old trees is not justification for killing a tree to improve a private view or to remove trees for urban infrastructure (that requires a whole different set of ethical and social considerations). It should also provide no solace to those who kill a tree before its time, such as the deadly ringbarking of the Separation Tree in Melbourne's Botanic Garden in 2010 and 2013.

My message at the Tree Forum was to get the very best expert advice on the health and longevity of the tree, talk to the community, make your decision, then explain that decision well and often. What could possibly go wrong?

The Australian, May 2015[25]

Postscript

My presentation to the forum was well received but, as I flagged, many more trees of a certain age have been removed and replaced across Sydney parks in the intervening decade. The event was also held in Melbourne, to a sympathetic audience. I recall far more tension at a Sydney tree forum run by the Australian Institute of Landscape Architects in May 2005, less than a year after the fig tree removals. Time and reality heals.

The multicultural garden

Create a garden that brings the world together

From the deck of my home, I overlook one of Sydney's spectacular, but fire-prone, sandstone escarpments. Within minutes I can enter one of the world-renowned national parks that weave their way through suburban Sydney. Our garden includes a mix of plants, some bequeathed with the house, others planted by us.

There are species from the local area, from wider New South Wales, some from tropical rainforests in northern Australia, and yet others from far-off countries such as China and North America. In the evenings, the local lorikeets congregate in the garden. A few weeks ago they harassed our almond tree, but more recently they have selected a bottlebrush from Western Australia to harangue. There are lots of other birds and reptiles, and a few mammals – egg-laying and not.

Earlier this year, with that scene in front of me, I read in Tim Flannery's Australia Day address that 'roses, lawns and plane trees are a blot on our landscape', implying that misguided gardeners should 'go Australian'. The follow-up on talk-back radio and in newspaper letter columns was all about growing native plants to help the

environment, and particularly to attract native animals. Flannery made some good points about water use and the need for a careful consideration of our environment when planning a garden, but does this really mean we should grub out our roses and replace them with Australian natives?

There are good arguments for planting indigenous species in our gardens if we wish to encourage local wildlife and to recreate the environment that existed before our particular housing development took root. There are fewer arguments for planting species that happen to grow somewhere in Australia but are not native to our particular area. Put simply, in Sydney there is little reason other than national pride to plant a Western Australian eucalypt in preference to a species from, say, one our near Asian neighbours.

To take this a step further, if your garden is in an urbanised area and the plants you grow do not escape into any nearby bushland, you should feel comfortable planting whatever you like.[26] Our houses are not caves in natural rock: they are human constructs in a very human environment – artifices made from wood, brick, bitumen and concrete.

One good argument for planting indigenous plants is to enjoy and assist the associated critters that might be attracted to such an environment, noting that these animals will be living in a mixed landscape, quite different to their traditional home. We mostly plant shrubs and trees, rather than a complex mix of annuals, herbs, fungi, algae and so on. This may not be a problem for them as individuals, but it could separate them from the rest of their habitat and cohort. Recreating a 'natural' landscape with its fully biodiversity is extremely difficult, if not impossible, and seldom what we want to achieve in a 'garden'.

A more subtle problem in planting local is the potential for genetic admixture.[27] Planting species that grow locally but are not sourced from your property or nearby could mix up the gene pool in an undesirable way. In this sense it can be argued it is better to grow something as distantly related and from as far away as possible – making sure it can't outcompete or interbreed with the local species, if, of course, that plant from further afield will not harm the environment.[28] Acknowledging that while planting Australian (or for that matter non-Australian) plants native outside your region you might attract some of the same animal species, or others, you have created a zoo-like environment, an artificial construct to favour certain wildlife, if not to constrain them. Again, this is fine, but one should not feel that it has some higher conservation value (other than perhaps for a charismatic animal species or two).

Even the water conservation argument, legitimately used to encourage native grasses acclimatised to the local area over common water-hogging Northern Hemisphere species, applies to individual plant rather than to its country of origin. Compare, for example, an Australian tropical rainforest garden to one including drought-tolerant plants of South Africa or the Americas.

Gardens do have values other than conservation and natural history. They are part of our heritage – just like our artificial construct, the house. The vast majority of people enjoy their gardens, no matter where the plants once grew. Many even appreciate having plants from other countries, with their related stories of discovery or associations with lives elsewhere. This is a sincere and reasonable human desire, to be encouraged, perhaps under the banner of the 'multicultural garden'.

To take an abstruse scientific example, recent DNA research shows that the North American plane tree is the closest relative of the Australian-rich family Proteaceae. One could argue that planting banksias and planes together reflects ancient evolutionary relationships. Of course, banksias and eucalypts together mimic present-day ecology, but only if they are species from the same habitat.

A simplistic call for gardens to be all-Australian risks alienating people who for whatever reason – abstruse or otherwise – chose the multicultural garden but also support genuine environmental initiatives.

The Gardens, June 2002[29]

Postscript

My view on overzealous advocacy for any particular planting palette hasn't changed since 2002, but my own plant preferences have. As they should. While (despite all my protestations) I still favour and plant more species from the land included within the Commonwealth of Australia, that's a personal and perhaps patriotic thing.[30] I've grown fonder over the decades of Australian garden stalwarts such as geraniums, common viburnums, clivias and pittosporums.

And camellias. Indeed, I've said that if it were within my remit, I would give honorary Australian Plant status to camellias. They are native to eastern and southern Asia, and most closely associated with China where many Australian camellia enthusiasts have rummaged in search of new species and forms. But Australians have bred camellias since the mid-19th century with 1,000 or so of the world's over 20,000 cultivars originating here.

Harking back to the Guiltfree planting principles at the start of this chapter, camellias don't escape into bushland and are pretty drought tolerant once established. They don't suffer from too many pests and diseases, and although most benefit from some protection on hot summer days, we mollycoddle them more than they need. It just so happens they didn't evolve in Australia.

Now there will be some who don't like camellias, and viburnums, clivias and pittosporums. A nod here to my friend Clive Blazey, of The Digger's Club, who feels that way about plants ranging from eucalypts (to be fair, in certain settings) to rhododendrons. While I agree with Clive that rhododendrons, like camellias, may have been over-used in gardens, that doesn't make them intrinsically unattractive plants. Some are less beautiful, to my tastes, but I'm willing to look past those to the cultivars or species that charm.

Yes, I'm judgemental of Clive, or anyone else, with strong views about particular plants. And I'll never like agapanthus, no matter what variety and no matter how unlikely some are to escape from the garden. Be as prejudiced as you want but don't present your personal inclinations as something of deeper and abiding significance. That, and agapanthus, I just won't allow.

Patriotism in the Victorian garden

Is it any better to grow plants from your own State than from elsewhere in Australia?

One midsummer's eve in 1910, Mr Pitcher presented a paper in praise of Victorian plants at a meeting of the Field Naturalists' Club of Victoria.[31] In the style of the time, he ended with a wordy and self-effacing apology: 'This paper may not have been very entertaining, but if it helps in achieving the objects mentioned I shall feel amply rewarded for the labour expended on its preparation.'

The objectives were three: to honour and preserve the few indigenous plants still remaining in the Melbourne's Botanic Gardens; to list the native Victorian plants growing in the Botanic Gardens; and

to suggest Victorian plants worth growing in gardens in and near Melbourne. Today we would warmly welcome the first and be thankful for the second. For the third we might ask Mr Pitcher to narrow and sharpen his focus.

Victoria is a political construct that in no way matches any natural plant distribution or habitat. Where Mr Pitcher pleaded 'not to overlook the Victorian flora when selecting plants for beautifying and decorating their home surroundings' we would shake our head and shrug our shoulders, perhaps whispering to our neighbour something about local provenance and genetic drift.[32]

On the positive side Mr Pitcher was recommending Victorian rather than Australian plants. This was progressive back then, and perhaps we regressed in the late 20th century when the talk was all about planting things native to the continent or nation of Australia.[33]

Pitcher begins his pitch by praising various parks around Melbourne that retain the 'character of the natural vegetation'. The remnant vegetation in the Botanic Gardens in 1910 consisted of a clump of swamp tea-tree (*Melaleuca ericifolia*) on Long Island, a manna gum (*Eucalyptus viminalis*) near the Herbarium and three river red gums (*Eucalyptus camaldulensis*): two on Princes Lawn near where 30 years earlier monkeys and birds had been caged in small zoo, and another at the bottom of Tennyson Lawn in the north-west of the gardens.

That third gum is the Separation Tree, named in honour of a gathering under its boughs on 15 November 1850 to hear Superintendent Charles Latrobe announce the passing of the Act to separate colony of Victoria from New South Wales (coming into effect on 1 July 1851).

A century of so on from Pitcher's address, there are still connections to the vegetation existing before the botanic gardens was established in 1846. A strip between Ornamental Lake and the Yarra River called Long Island has been replanted with indigenous vegetation including swamp tea-trees, although the original trees of this species are long gone (a swamp tea-tree on nearby Baker Island may be original).

The manna gum is alas no more. Two river red gums remain: the Separation Tree, now estimated to be over 400 years old, and one beside the Ornamental Lake called 'The Lions Head' (after a large gall near its base) that surely predates the botanic gardens.

The Separation Tree has had a rough time of late. Two attempts have been made in the last 3 years to ringbark it, leaving only a small fragment of cambium to deliver essential sugars to the root system. Despite this vandalism the tree looks healthier than in the photograph in Mr Pitcher's article. How long it survives now will depend to some degree on how much food has been stored in the roots and whether we can protect it from the next drought or disease outbreak – any stress is likely to do it in.

The other remnant river red gums are hanging in there, but are in the twilight of their lives. Offspring of the Separation Tree are thriving nearby, as are planted, self-seeding and coppiced examples of other indigenous species.

Mr Pitcher clearly felt more could be done to promote the growing of Victorian plants. He suggested a larger area at the Melbourne Gardens be devoted to Victorian species, that the 400 Victorian species already in the collection be clearly tagged (with 'a specially tinted label'), or that a new area be set aside exclusively for Victorian plants.

While there are no tinted labels, 1,058 of the 8,000 or so species grown in Royal Botanic Gardens Melbourne today are Victorian natives (with more in the Australian Garden and bushland at Royal Botanic Gardens Cranbourne). In addition to Long Island, featuring local plants, there are the five island garden beds devoted to Victorian rare and threatened species.[34] However, I would remind readers, and Mr Pitcher too if he was here, that Royal Botanic Gardens Melbourne is deliberately a 'world garden', displaying plants from nearly every country on Earth.[35]

I was curious to discover which Victorian plants Mr Pitcher would recommend for Victorian gardens, particularly for the area around Melbourne. He advocates 60 plants in 10 categories: wattles, eucalypts, ornamental trees, ornamental tall shrubs, ornamental flowering shrubs, smaller flowering shrubs, ground covers, creepers and climbers, hedge plants and hardy ferns. I shared his list with colleagues at Royal Botanic Gardens Victoria, asking them to identify any problem species and striking omissions.

A few plants are not actually Victorian natives, so they are easily excluded. Sweet pittosporum (*Pittosporum undulatum*) and coast tea tree (*Leptospermum laevigatum*) were rejected by my reviewers as too weedy. A few others were considered rather difficult to propagate.

It was noted that the list omitted some plants very popular today: for example, silver wattle (*Acacia dealbata*), blackwood (*Acacia melanoxylon*), red box (*Eucalyptus polyanthemos*), snow gum (*Eucalyptus pauciflora*) and all correas. Shockingly by today's standards, he included only one *Grevillea* (*Grevillea lanigera* listed as *Grevillia ericifolia*).

All in all, though, his list offers a reasonable choice for gardens around Melbourne. It includes bottle-brushes, heaths, westringias, running postman (*Kennedia prostrata*), two *Clematis*, and two tree ferns.

Here in 2015, growing Victorian plants doesn't make a lot of sense to me. I can understand 'Australian plants' as a theme, as long as it's done for cultural and aesthetic reasons rather than in any sense of connection with our particular chunk of this vast and diverse continent.

I can also understand growing local indigenous plants, although there will always be some debate about how local and how indigenous (they may be pre-European, pre-Aboriginal, naturally invasive, dependent on humans, and so on) and the matter of interbreeding with the 'real' local plants.

One could argue for a regional focus, such as one of the Natural Regions of Victoria (championed in the *Flora of Victoria*).[36] Metropolitan Melbourne is carved up among the Gippsland Plain, Midlands, Eastern Highlands and Victorian Volcanic Plain. Perhaps habitats, such as alpine, rainforest or wetlands, would work? But then why not include a selection of the world's plants that thrive under such conditions, as long as they meet sensible planting criteria?

In the end I'm drawn back, as always, to planting what you like as long as it has a neutral or net benefit to the environment. If you like native creatures or, quite rightly, believe we should encourage their abundance and welfare, you might grow a selection of plants from the local area, but equally from Australia and from overseas, that favours them. You must, of course avoid seriously weedy plants and those needing environmentally deleterious chemicals to save them from local pests and diseases. From this reduced list you can choose the plants that make you smile, relax or write poetry, depending on what it is that you get from your garden.

Mr Pitcher might not be happy with my drift away from Victorian plants, but we are a demanding lot, forever seeking new variety and stimulation. The 3,700 or so species of plant in this State were never going to be enough to hold our attention, particularly, as Pitcher admits, 'it has to be borne in mind that a large number of our indigenous [Victorian] plants...are too insignificant for cultivation, and are of no popular interest whatever, while many others it would be either impossible or undesirable to grow...'.

Why should we resist the chance to grow Western Australian banksias and Sydney's Wollemi pine, or proteas from South Africa and camellias from China? As Paul Keating put it back in 2007,[37] referring to the then Prime Minister John Howard and borrowing from George Orwell, 'for him, Gallipoli was an exercise in nationalism. For me, Kokoda was an exercise in patriotism.'

Keating went on to explain that 'a patriot will not exclude a person from the community where they have lived side-by-side and whom he has known for many years, but a nationalist will always remain suspicious of someone who does not seem to belong to his kind of people, or more likely, his kind of thinking'.

Putting politics aside, I have no problem with patriotism in the garden but little time for nationalism.

Wildlife Australia, summer 2013; The Age, January 2015[38]

Postscript

Ten years later, my perspective on encouraging Victorian plants into our gardens has evolved a little. In fact, I recently wrote an essay for a horticultural journal extolling the virtues of growing more rare and threatened Victorian plants in Victorian gardens – largely for their value as an educational and promotional tool, but also perhaps as part of a broader 'distributed' conservation collection (although that is far more complicated).[39]

The Separation Tree is sadly no more, pronounced dead in May 2015, a few months after this essay was published. It failed to recover from the two ringbarkings, despite efforts to bridge the gap with compresses of peat or grafted saplings. There are descendants growing nearby, propagated from seed of the tree, including one planted in 1951 to celebrate the centenary of the separation of Victoria and New South Wales. Other offspring have been distributed to schools around the State as part of a programme in partnership with the Victoria Day Council. All can become alternative places for reflection on this tree's life, from its enduring connection with the Wurundjeri Woi Wurrung and Bunarong peoples through to the creation of Victoria, and thereby a Victorian flora.

Top five 'first' botanic gardens

To find the oldest botanic garden in the world, first define what you mean

In a lengthy paper describing the origins of botanic gardens (and more), my colleagues from Royal Botanic Gardens Victoria, Roger Spencer and Rob Cross, stated that the 'oldest *existing* botanic gardens date back to the early modern period, to the educational physic gardens associated with the medical faculties of universities in 16th century Renaissance Italy' (my emphasis).[40]

THE DIRECTORS OF THE
WORLD'S BOTANIC GARDENS
RESOLVE WHICH GARDEN
IS THE OLDEST.

This is generally accepted, although there is some debate – alluded to in Spencer and Cross' article – around which of those early Italian botanic gardens is the earliest. Also debated, by some authors, is whether the agricultural gardens established in Moorish Spain during the 11th century might be considered botanical in the sense I define it later.[41]

There is a minor quibble about continuity. The first garden we recognise as botanic was Orto Botanico dell'Università di Pisa, commissioned by Cosimo I de' Medici and constructed in 1543 and 1544 beside the River Arno in Pisa. This garden for *simplicia* (medicinal plants) – what the English would call a physic garden – was relocated in 1563 and then 1591. It resides today only a few blocks away from where it was first established, but these moves relinquish its title as the oldest botanic gardens still *existing* and *persisting* (in its original location).

That honour has been bestowed on Orto Botanico di Padova, completed in early 1545. Also a garden for *simplicia*, it sits today where it did in 1545, opposite the 14th century Basilica of Saint Anthony in Padua (Padova), near Venice, on the opposite side of the Italian Peninsula. Orto Botanico di Padova narrowly beats Orto Botanico di Firenze, in Florence, 70 kilometres to the west of Pisa, which opened in December of the same year. So, gold to Padua, silver to Florence and bronze to ... No, not Pisa, and not one of the Italian city states.

Before we leave Italy, and the Mediterranean region, you might wonder whether there were botanic gardens, or things like them, in antiquity. This is a far more interesting question than whether a botanic garden needs to stay fixed to the same point on the Earth, but one still requiring some definitional dexterity.

The closest match to something we might call a botanic garden today was, as Spencer and Cross observe, a 4th century BCE garden at the Lyceum in Athens. It certainly contained a collection of plants used at least in part for scientific observation. To my knowledge the plants were not labelled in any way, a characteristic of every botanic garden today, and the landscape and horticulture were probably not particularly 'ornamental'. While we may focus on the collections and the science, Spencer and Cross quite rightly argue that contemporary botanic gardens generally include ornamental horticulture as part of their 'mix of science and education, art and utility'.

The first botanic garden to be convincingly ornamental *and* scientific was established in the Dutch university town of Leiden, in 1587. I am going to also award it the bronze medal for oldest botanic garden in its original location, despite the distinct possibility of other smaller botanic gardens emerging on the Italian peninsular between 1545 and 1587. Hortus Botanicus Leiden was perhaps the first of these early *botanical* gardens to offer more than an orderly display of plants for medicinal use. More than a physic (or simplicia) garden.

While the desire for a garden to combine both science and ornament was a European one, ornamental gardens and gardens for contemplation were of course a more universal concept. If we allow a garden of medicinal and useful plants to be a botanic garden, then I'm sure many gardens in Asia would not only qualify but precede those of Italy. In any case, botanic gardens of the medicinal kind found in Europe began to appear in the distant colonies established by European nations during the 16th century.

This includes a botanic garden in the Portuguese territory of Goa, in what is now India. However, the title of earliest botanic garden in India is usually attributed to a garden established by an East India Company officer in Kolkata, two centuries later.[42] Established in 1787, the Royal Botanic Garden, Calcutta became more recently the Acharya Jagadish Chandra Bose Indian Botanic Garden, named in honour of a Bengali botanist-physicist and an early writer of science fiction.

By the early 17th century, the concept had crossed the Channel, with the University of Oxford creating its first botanic garden in 1621. It took until 1673 for England to get its second botanic garden, albeit one with a particular purpose as its name makes clear, the Chelsea Physic Garden. The Royal Garden at Kew was begun by Princess Augusta (mother of George III, who later lived there) in 1759.

Back on the continent, in 1718, Jardin de Plantes in Paris changed its emphasis, and name, from one devoted to medicinal plants to plants more generally. In the United States, the Missouri Botanical Garden in St Louis, established in 1859, describes itself as 'the nation's oldest botanical garden in continuous operation'. So too does the US Botanic Garden, in Washington DC, which opened in 1850 – 'the oldest continuously operated botanical garden'. I should also mention Bartam's Garden in Philadelphia, established in 1728, self-described as 'the oldest surviving botanic garden in North America', but I'm not convinced.

What about in the Southern Hemisphere? Back when I was worked at Sydney's botanic gardens – and for reasons I can longer remember –

I would say that the Jardim Botânico do Rio de Janeiro in Brazil, established in 1808, is the oldest south of the equator, followed by gardens Sydney in Australia (1816) and in Bogor in Indonesia (1817).

I may be being a little parochial, having visited all three and being director of the middle one. In any case, Spencer and Cross would award the southern gold medal to Sir Seewoosagur Ramgoolam Botanical Garden in Mauritius, established in 1736, or more formally as a botanic garden in 1767. Then there is the Company Garden in Cape Town, begun in 1652 and already, according to Spencer and Cross, 'an exceptional botanic garden' by 1680. I visited the Company Garden in 2005, now outside the botanic garden network in South Africa. I reckon it can be considered a botanic garden still, albeit not an exceptional one.

As you will have gathered by now, it's a vexed business tracking down when a garden begins, and even harder, perhaps, to determine when it starts being to be a botanic garden. All botanic gardens will be a little less 'botanic' in their early years even if established as such right from the start. The mobility of some makes it even more difficult to nail down when and where they start to function as a botanic garden. Then there is the fundamental question of what makes a garden botanic. Many have attempted a definition but let's focus on the last 50 years or so.

In 1963, the International Association of Botanic Gardens, of which I was President from 2017 to 2025, defined a botanic garden rather succinctly as a garden 'open to the public and in which the plants are labelled'.[43] Short, sweet, but rather inadequate. While it excludes any public garden without the names of at least some of its plants visible or searchable in some way, it includes any public garden that choses to do just that.

The next major attempt to nail down a definition went to the other extreme. Authored by Vernon Heywood, then Director of Botanic Gardens Conservation International, a 1987 offshoot of the International Association of Botanic Gardens, 'The Botanic Gardens

Conservation Strategy' lists 11 expected attributes of botanic garden.[44] These attributes include being open to the public and (adequate) labelling of plants but extend to scientific research, communicating information, seed exchange, documented plant collections and promoting conservation. The authors stress that such a list 'does not constitute a comprehensive summary of the activities undertaken by botanic gardens'.

In 2000, Botanic Gardens Conservation International published an update of this document, called the 'International Agenda for Botanic Gardens in Conservation'.[45] Here, the definition is more one of intent rather than trait based: 'institutions holding documented collections of living plants for the purposes of scientific research, conservation, display and education'. This is a far better approach, allowing botanic gardens to look and operate in a way that suits their local environment, but sharing some common goals.

While I like this definition, I felt I could improve on it a little, presenting the following to the 6th Global Botanic Gardens Congress in Geneva in 2017. I argued that historically, botanic gardens were in their fourth generation and we should expect them to be 'a scientifically managed and inspiring landscape of documented plant collections, where every plant and setting has a purpose'.

More importantly for my purposes here, do any of these definitions help to create our chronological list? In common, they all call for *documented*, or slightly more limited *labelled*, collections. Being *open (and welcoming?) to the public* is considered essential, although implied rather than explicit in the 2000 and 2017 definitions. Apart from the 1963 definition, the plants must have some acknowledged *purpose*. The role of *ornamental horticulture* comes through in the last two definitions: in 2000 as *display*, and in 2017 in the reference to *inspiring landscape* plus the addition of the word *setting* as well as plants when it comes to purpose.

While I obviously favour my 2017 definition for a modern botanic garden, I don't see it as an exclusive one. Perhaps aspirational for some.

If I had to list some fundamental attributes to accompany this definition, they would be:

- A documented collection of plants (to some purpose)
- An ornamental and purposeful landscape
- Open and welcoming to the public
- A scientific (or if you prefer, evidence-based) approach to their management and use (noting that this doesn't necessarily require an active scientific research programme by the organisation itself)

These points will not please everyone (and I'm sure they could be improved) but using them, albeit subjectively, let me return to my list of firsts, or nearly firsts, acknowledging that being first says nothing about the quality of the botanic garden, then or now. And I've decided to adopt the need for the botanic garden to stay in the same place, more or less (i.e. with at least some common ground throughout its history). Otherwise, we would need to evaluate the scope and concept of the organisation running the garden over its lifetime, and whether the subsequent gardens carried through some core element of the previous one. All of this I consider too difficult.

As to botanic gardens that existed once but not now, they are simply too hard to document and review (although such a list would, I admit, have validity equal to mine). The nomenclature for the organisation, the city and the country are as used today, with the geographic location in English.

So, here is my list of the top five 'First' botanic gardens, with apologies to all the countries and regions left out (the regions mentioned are those I've been asked about recently).

First on Earth
1545: Orto Botanico di Padova; Padua, Italy
First on Earth with strong scientific and ornamental values
1587: Hortus Botanicus Leiden; Leiden, The Netherlands

First in the United Kingdom
1621: University of Oxford Botanic Garden; Oxford, United Kingdom
First in the Southern Hemisphere
1652: The Company Garden; Cape Town, South Africa
First in the Americas
1850: US Botanic Garden; Washington, DC, United States

Talking Plants, March 2018[46]

Postscript

I wrote the original version of this essay after 5 years as Director and Chief Executive of Royal Botanic Gardens Victoria, following 2 years working at Royal Botanic Gardens Kew, and before that, 8 years as Executive Director of the Royal Botanic Gardens and Domain Trust. The Trust's responsibilities included the Royal Botanic Garden Sydney, which began life in 1816, making it the first in Australia. We didn't talk much about Sydney's antecedence at Royal Botanic Gardens Victoria, which was established in 1846. We did, however, take pride in Melbourne's Garden being established only 11 years after the city itself. In Sydney, it took 30 years.

I received a few suggested additions in response to this blog post but none that would change the 'Top Ten' – other than of course including categories for other regions and countries. One of interest is St Vincent and Grenadines Botanical Gardens in Kingston, established in 1765 and described as 'the oldest in the New World'[47] or sometimes the oldest in the Western Hemisphere or in the tropical world. Regarding my list, it is in the Northern Hemisphere so preceded by those 16th and 17th European botanic gardens, but it is the first in the Americas (aka Western Hemisphere), jettisoning the US Botanic Garden.

As to the botanic gardens of the Old World outside Europe – a major omission from my essay – the Sir Seewoosagur Ramgoolam

Botanical Garden in Mauritius (1736 or 1767) might be considered the oldest in this category. Followed, perhaps, by the Acharya Jagadish Chandra Bose Indian Botanic Garden in Kolkata, from 1787. In China, botanic gardens arrived with the colonialists from England, with earliest the Hong Kong Zoological and Botanical Park, opened in 1860, followed by the Nanjing Botanical Garden Memorial Sun Yat-Sen, in 1929; today there are more than a 150 botanic gardens in China, most of them created in the last few decades.

I include some of this reasoning and chronology in my recent memoir,[48] as a prelude to locating the world's best botanic gardens. That quest became a major theme of the book.

For pleasure and ornament

> *Gardens are places where plants are cultivated primarily for pleasure or ornament*

In my closing remarks at the 2022 National Conference of the Australian Garden History Society, I observed that this was a 'knowledge society', one that generates, shares and applies knowledge to foster human development. At least that's how I was quoted in an editorial in the Society's magazine, *Australian Garden History*, soon after.[49]

The convenor of the Hobart conference, Prue Slatyer, asked in her editorial how the society might strengthen its role in support of this intent. The answer emerging from the conference, wrote Prue, was to broaden its scope beyond gardens to landscapes more generally. As a provocation in my remarks, I asked when I might be able to introduce the society as 'the Cultural Landscape Society of Australia'. Was this a better name for a society with a mission to 'promote awareness and conservation of significant gardens and cultural landscapes'?

Prue Slatyer mused a little further on what that change might mean, including a broader cultural as well as spatial perspective.

She suggested better integration of knowledge from Aboriginal people, a continued focus on climate change (at that same meeting I launched the society's Climate Change Position Statement) and more advocacy on behalf of significant gardens and cultural landscapes.

With time to cogitate further, my thinking has changed on both the name of the society and its scope. Not that I was even then really advocating for CLSA to replace AGHS! However, I do think it is time to re-examine the society's nomenclature.

Let's start with the etymology of the word 'garden'. I note, as does horticultural taxonomist and historian, Roger Spencer, in his essay 'What is a garden?',[50] that most derivations include an enclosure, or as Roger puts it, a 'bounded space'. Roger skirts around our need and indeed our ability to define 'garden' more precisely, but considers in addition to a bounded space a requirement for cultivated plants to be present and managed, and for the garden to be usually associated with some dwelling such as a house.

This is close to where most dictionaries land. Common to most on the internet (Cambridge, Oxford and Collins) and my own printed copy of the *Shorter Oxford English Dictionary*, a garden is a piece of land near a house where flowers and various other plant life are grown. It may also be a public park with similar ornaments, such as a botanic garden.

Dr Samuel Johnson[51] favoured 'a piece of ground enclosed, and cultivated with extraordinary care, planted with herbs or fruits for food, or laid out for pleasure'. I like his typical flourish about extraordinary care but more so his desire to define a purpose to the garden, and particularly the final option, 'laid out for pleasure' (noting that a commercial orchard today would fall within Johnson's other purpose). As with my attempts to diagnose a 'botanic garden' over the years,[52] I prefer definitions based on intent or purpose, rather than traits.

Accordingly, my submission is: *gardens are places where plants are cultivated primarily for pleasure or ornament*. This intent would exclude things that most people would not consider gardens, such as landscapes managed primarily for pasture or agricultural crops, ecosystems restored or regenerated, and large tracts of land where human impacts are few. Although excluded from the definition, these may be topics to explore to improve our understanding of gardens themselves.

I worry about the term 'cultural landscape' because on further reflection I think all landscapes are cultural in some way, and even more narrowly, all landscapes are manipulated deliberately or indirectly by human culture. This may be weed control and cultural or prescribed burning at one extreme, to inadvertent bushfires and weed introductions at the other. It also includes human impacts on the climate.

I've used the word 'cultivated' as a softer (and therefore more acceptable?) term than 'managed' or 'manipulated', but they are much the same thing. Other words such as 'weed' and 'prescribed' are with used trepidation but for convenience. The terms 'pleasure' and 'ornament' are equally fraught, but I have in mind something similar to what British garden writer Edward Hyams observed (as quoted by Roger in his essay), that gardens are 'surplus to necessity'. Not that they don't enrich and save lives, but they are, by and large, something we pursue beyond mere survival.

If a garden is a place where plants are cultivated primarily for pleasure or ornament, does that work for the Australian Garden

History Society? My understanding is that it was established in 1980 to bring attention to the planted landscapes of larger properties and estates, and to appreciate, promote and care for these significant creations. My definition includes them. It also includes smaller home gardens, a herb garden created for effect as well as produce, a balcony ensemble of pot plants, and a raked sand garden as long as there is a pine tree at one end.

While an orchard, canola crop or river red gum-spotted grazing land may be intensely pleasurable, I would say my definition excludes them. Similarly left out are forests, grasslands and heath that may look like they do thanks to management by humans over centuries or millennia – again acknowledging they can bring great pleasure and ornament.

None of this is to exclude such topics for the purview of the society, just not to have them as its primary purpose. Otherwise, it becomes a society for everything, which would result (I think) in a society for no-one. Rather than call for us to become a society for 'cultural landscapes', I rather like the conceit of a knowledge society for gardens.

We should make at least a few tweaks to nomenclature. I was going to suggest that we simplify the society's mission by removing the arguably superfluous term 'cultural landscapes', given all I've written above. However, these words were added quite recently with careful and worthy intent, and the word 'landscapes' warrants a place somewhere in our societal descriptors. It is worth thinking about whether 'cultural' is the right adjective (or indeed if any adjective is needed) preceding 'landscapes', but for now let's leave the mission intact.

More importantly, I think the word 'history' is too restrictive for all that we do. I'm keen on history myself but I don't think it needs to be in the name of the society. The important ingredients for me are 'Australia' and 'Garden'. Which led to my first attempt at a new name, 'Australian Garden Society'. I shared this with the National

Management Committee of the AGHS and others prior to them meeting in Melbourne in February 2024.

Between sharing and the meeting, I was discussing the matter with my close confidant, and also deputy-chair of the Victorian branch and AGHS (Vic) newsletter editor, Lynda Entwisle. She suggested adding the word 'heritage', which includes concepts of conservation, preservation and caring for places. All very apt. That led us to 'Garden Heritage Australia', ridding our name of the word 'society' and 'history', adding in the well-understood term 'heritage', and giving the whole thing a bit of pizzazz by reordering the words to make it sound more contemporary.

With either Australian Garden Society or Garden Heritage Australia – and I'm now favouring the latter – the Society's current mission statement works well (with some minor tweaking if desired) and the journal could keep its name *Australian Garden History* (to create a point of difference with the society name and to make it easier for librarians!).

Elsewhere, we might strengthen any statements about the society to make it clear we are interested in gardens big and small, and that our definition of garden is broad and inclusive. A pithier statement (a byline or slogan) could be drawn from the mission to make it abundantly clear that we include gardens and other 'constructed living landscapes' (and these are *not* the right words). We remain a society but don't need to go on about it in our name.

I'm comfortable that in both short-listed titles, the word 'garden' refers to places rather than 'gardening' as a practice. How we garden will inevitably form part of the society's consideration, but it is not its primary purpose. Similarly, our inevitable interest in individual plants should accompany rather than dominate our intellectual interest in the history, management and care of places where plants are cultivated primarily for pleasure and ornament.

Let me leave it here, ripe and ready for debate, and before I change my mind again.

Australian Garden History Society, April 2024[53]

Gardening in a post-modern world

Why I believe in an interventionist gardener

One fine spring morning in 2015, I met landscape designer and writer Georgina Reid in the Camperdown Cemetery in Newtown, a gritty, inner-west suburb of Sydney. I was interviewing Georgina for *Talking Plants*, a radio show I co-hosted over two summers with garden designer Jim Fogarty on ABC RN.

Georgina wanted to show me how this cemetery was different to most, and why it was a special place for her. Unlike so-called memorial parks, Georgina explained, this cemetery has a genuine park-like feel about it, partly through age and (perhaps benign) neglect, partly through design. 'It's a bit rough', says Georgina, 'not manicured in the slightest, which appeals to me.'

The grass was patchy and rarely mown, tufting up around the green-algal stained headstones. Most of which had a jaunty lean. Weedy creepers nestled beside indigenous remnants or plantings, while mature figs, eucalypts and oaks provided plentiful shade for the dead and the living. Among the living that day were volunteers weeding and supplementing the understorey with plants of local provenance. Right here is the contradiction behind what I'm going to call 'non-gardening' or 'gardening lite', terms Georgina wouldn't use but which are close to the philosophy of gardening on which she writes and speaks today.

A few years after my cemetery visit, horticultural journalist Megan Backhouse summarised Georgina's approach to gardening in her column for *The Age*: 'gardens are not objects', 'question everything' and 'all plants are equal'. The supplementation of the cemetery landscape – now a parkland of exotic and local trees with patches of indigenous grassland – was consistent with their views on a changing landscape, one not defined by a particular kind of plant, or planting style.

My own take on this 'anything goes' approach is 'as long as you like it and you don't destroy the world in the process'. I'm happy with some bold streaks of local plants if that makes the landscape more interesting, beautiful or functional (which might include encouraging local wildlife).

Georgina has recently championed a lighter touch when it comes to landscape design. She gave a passionate and provocative speech at the Australian Landscape Conference held in Melbourne in 2023, advocating more garden-mending than garden-making. Essentially, less is more and if we do intercede, says Georgina, it must be with due attention, care and humility. The term 'reciprocity' was used a number of times.

Gardening, for Georgina, is about asking questions rather than following garden design rules. Those questions are very much about the place, its existing plants and all that goes with these. For Georgina, an effective garden is one that respects what was there before the garden designer or gardener stepped in. Take time to look and listen, and to understand, before turning the first sod – if indeed there is a need to turn any sods.

I agree with this as a starting point but once you've done that, I reckon you can create what you want to; again, with due deference to what is going on *outside* your garden cocoon. The best gardens will bring out the intrinsic values of the site itself but perhaps not always – it might be something simply of itself.

This may be where we part ways. Georgina isn't thrilled by the consequences of Western thinking and science, and of course she has a point about the many things we have stuffed up through lack of empathy with the natural world. I see life, for better or worse, through the stories of Western science. Hopefully, this sits alongside some compassion, tolerance and space for other world views, but I am a particularly rational being.

Georgina starts with beauty and contemplation (and awe, which we'll get to), and then moves on to repair if needed. An ideal gardener for Georgina is someone who mends, someone who reads Country to find out what it wants, not what the designer wishes to impose on it. Focusing on what is present is about not treating the site as a blank canvas.

I think the garden repair approach comes from gardening, as Georgina is doing in her current home garden on the Hawkesbury River, a place still part of a bigger, more intact natural environment, not fully embedded within suburbia. But I have no problem with creating a better or new garden in urban environments like the inner west of Sydney when that land is adjacent to a constructed house facing onto a bitumen road. That's the borrowed landscape.

At the end of her talk, Georgina praised a sense of awe and wonder in a garden and I'm all for that. Perhaps, I thought, we are not so far apart in our thinking. It might be through intensifying the inherent values in a place or, from my perspective, by redesigning and creating something afresh. Keep connected, certainly, but don't let that constrain creativity. Our differences may stem from how we both view gardens and gardening. Georgina defines gardening as 'a framework for being with the world' and a garden as 'any place where human hands and feet touch soil'. My definitions are far narrower and more pragmatic – more Western, I suspect.

You'll have gathered from the previous essay[54] that I see cultivation (or even more loaded words such as 'management' and 'manipulation') as an essential part of a garden. If you don't cultivate, it isn't a garden – which is fine, of course. After you commune with the place and its plants, you may decide that it is pretty enough or would not benefit from your intervention. If it is what many would call a natural landscape, then turning it into a garden may be inappropriate and just stupid.

If you decide to garden, then like your house, your clothes, your writings or art – your entire life – it can be what you want it to be. Gardening is most often the deliberate introduction of plants into a place where they do not grow naturally or in quantities and arrangements different to how they might grow in their natural state. That's the whole point of most gardens. They are a human cultural construct, based on the desire for pleasure or ornament, as I argued previously – and they are not intrinsically better or worse than nature for all that.

While I can appreciate a garden returning to nature with varying degrees of assistance from humans, I don't place any greater value on that outcome above any other. I value complex ecosystems where the human touch has been light, and will support their protection in reserves, covenants and the like. But a garden is a thing like a gallery, a museum, a sculpture or a football game. It is a construct to be enjoyed because of its artificiality, even if it sits within some other less constructed system.

This becomes important when we consider what to do with garden landscapes separated from their designer – typically by death. Of course, you can do what you like, but there is nothing wrong with taking a particular position or approach based on your interpretation of the creator's intent (or not). The fear of somehow destroying the creator's vision inhibits some people from intervening more actively, but gardening around the edges is still gardening, in the same way you might care for a heritage garden based on a clearly established design.

As Georgina and I both understand it, a garden is not an object. It is a changing thing that demands we continue to intervene creatively. Adding new garden beds to heritage gardens, inserting native grasslands into an otherwise ornamental cemetery woodland, and replacing fallen or poorly performing trees with ones better suited to the climate or – even better – to the design are all reasonable and laudable interventions.

Gardening is a creative and active pursuit. We can make gardens that connect to the place or to another time or to some other insight. Gardening, like art in a post-modern world, has no single or absolute truth, and we should be always ready to change our perspective along with its material elements.

The Museum, July 2024[55]

Postscript

I mentioned the artist and gardener Burle Marx in my introduction to this chapter in regard to how the garden he created near Rio de Janeiro, Brazil, should be managed following his death. Australian landscape architect, Julian Raxworthy, provides this and another example in his book *Overgrown*.[56] When British poet and gardener Geoffrey (G.F.) Dutton died, his 'marginal garden' (described as gardening around the edges) in Perthshire was allowed to effectively rewild by his family. Raxworthy asks whether that is the right or reasonable thing to do. Of course, gardeners can do what they like, particularly when you are landed this role through family or legal responsibility, but Raxworthy notes the garden is now 'going backward'. His advice to the family it to 'take a position'. I agree.

4

Gardening with convictions

At its essence, gardening is a straightforward pursuit. Simply place a plant suitable for your climate in the right kind of soil, with enough sunlight and make sure it gets the water and nutrients it needs. You can prune if you like.

The devil, of course, is in the detail. Which is what makes gardening interesting and different to most other things we construct on our block of land. Not only do you need to understand the intrinsic conditions of the place but you need to respond as each plant grows, flowers, fruits and eventually senesces.

Typically[1] our job as gardeners is to create a pleasing contrivance by manipulating the plants and their setting. Most of us interfere with nature in our garden, striving to protect the plants we want while eliminating those we don't. You can see from this language that gardening is a lot about control.

The 'organic gardener' still wishes to control their plants but without the use of chemicals and other processes that might harm the immediate or broader environment. That sounds like a worthy objective. But what if by doing this there are perverse outcomes such as needing to destroy 'natural' vegetation to create your farmland, or ineffective weed control reduces yields (requiring yet more land) or allows plants to spread beyond the garden? Everything we add to a garden is a chemical or mineral, so which are good and which are bad?

In the following essays, I consider gardening techniques and principles that are not quite what they seem. Seaweed supplements, companion planting and the wood-wide web don't live up to the hype (although the microbial world beneath are feet certainly is impressive).

I argue we should keep an open mind about genetic modification and glyphosate, but suggest you ignore the lunar cycle. Oh, and burying your dead in the garden is probably a good thing.

Converted to seaweed

You can add seaweed to your garden but don't expect miracles

Those who know me well would think I'd be in favour of spreading a bit of seaweed around the garden, combining as it does two of my professional passions: algae and plants. On the other hand, I might resist the killing and harvesting of marine algae – or seaweed as we call it disparagingly – preferring to see it live a natural life on our rocky shores.

Whatever my biases, I've always been a little sceptical of the benefits of adding seaweed to the home garden. That has now changed. A little desktop research, and a review of my prejudices, has made me if not an evangelist at least an advocate for seaweed-assisted horticulture.

Seaweed manure has been used in horticulture since early Roman times, including a rather specific recommendation that when

cabbages are transplanted – at the sixth leaf stage – their roots should be mulched and manured with seaweed. A few centuries later Romans recommended seaweed be added to the roots of pomegranate and citrus.

Seaweeds have also been a source of soda ash and potash since the 17th century, although production wound down after the First World War. Potash provides plants with one of three essential macronutrients, potassium, so the mode of action is clear.

Less so with the emergence of liquid seaweed extracts in the middle of the 20th century. These are said to 'stimulate' plant growth, which would seem to be a good thing but rather vaguely expressed.

My own doubts date back to 2006, when I read an article in the quarterly magazine of the Australian Skeptics, *The Skeptic*.[2] Retired chemical scientist and self-confessed avid grower of chrysanthemums, Geoff Sherrington, answered the question 'Can kelp work?' with an emphatic 'no'.

Kelp, a brown seaweed or alga, contains so little nitrogen, phosphorus or even potassium that, according to Sherrington, 'a farmer wanting to add nitrogen to a patch of land can add a tonne of urea or somewhere between 50 and 100 tonnes of kelp'. He highlights the trial against seaweed producer and seller Maxicrop, in New Zealand in the 1980s, a company that sued unsuccessfully for defamation, negligence and misfeasance. Contrary to the company's advertising, the recommended dose of seaweed extract *did not* boost plant growth.

Apparently growth stimulants (hormones) present in kelp are too low in concentration to have any beneficial effect, and seaweeds in general are not a particularly good source of minerals and other micronutrients. An article with a broader remit by agricultural scientist, David Conley, was published a few years earlier in *The Skeptic*.[3] He scoffed that one seaweed extract manufacturer boasted over 70 micronutrients in its seaweed product but failed to mention that only 16 were needed for plant growth. So through a mix of excess hyperbole and no demonstrated mechanism of action, it seemed seaweeds were doomed to become a fallacious fad.

But let's dive a little deeper. First up, could seaweed do any harm to your plants? Like anything that lives in the sea it contains a lot of sodium chloride, and anyone trying to garden by the coast will know that not all plants appreciate high levels of salt. So dumping large rotting heaps directly on the garden is unlikely to be a good thing. That said, adding a little dried seaweed (or seagrass) might provide useful mulch and help to reduce water loss and improve soil texture.

But it's liquid extract from kelp that has led the seaweed revival in recent years. Kelp is just one kind of marine alga. There is a vast variety of biology, chemistry and ancestry among what we call algae, far greater than we find in say flowering plants or mammals. (Seagrasses, by the way, are not algae: they are flowering plants that just happen to grow under the sea.)

Algae are very much part of our non-gardening lives. You might recognise them wrapped around sushi or floating in miso soup, but their extracts are also used to thicken toothpaste and ice-cream, and to create the glacé cherry. While it doesn't seem logical to add algae or seaweed to our garden soil to make it 'set', adding something that absorbs and holds water might a good thing.

Research published in 2009 stated clearly that 'seaweeds and seaweed extracts enhance soil health by improving moisture-holding capacity and by promoting the growth of beneficial soil microbes'.[4] However, while the improved plant growth was demonstrated in experimental trials, the cause of action was 'largely unknown'.

By 2011, researchers were comfortable stating that while 'the earlier concept that benefits of seaweeds and their extracts were due mainly to their manurial value or to their micronutrient suites is no longer tenable ... benefits derived from the use of seaweed products can be indisputable in agriculture'.[5]

Since that time there have been numerous positive responses in trials for germination, root development, quality of the leaves, general vigour of the plant, resistance to pathogens and even resistance to drought and frost. While seaweed itself may not be a great source of the major nutrients, adding an extract to the soil can increase the nutrient content in leaves. Presumably, something in the mix helps a plant gain better access to what is already available – maybe some of those thickeners break down to form what are called 'chelating agents'. Growth hormones are not, it seems, the key ingredient.

Research and interest continues, particularly around how seaweed extracts may help plants with nutrient deficiencies or to deal with stress and disease. Plenty of demonstrated benefits but still a poor understanding of what the seaweed brings to the problem. This accounts for the statement printed on at least one popular brand of seaweed extract product: that it is a 'plant conditioner'.

Meanwhile global production of seaweeds for soil and plant applications runs at more than 500,000 tonnes per annum. About 95 per cent of seaweed harvesting – for all purposes, including sushi and toothpaste – is from farmed algae, which is good if mangroves and other natural marine vegetation are not cleared to create the farm.[6] Wild collecting makes a small contribution and has to be managed carefully.

So where does that leave me, the sceptical botanist? Well, if seaweed can be harvested in a reasonably sustainable way *and* there is demonstrated evidence of it producing healthier plants – even if we don't quite know the reason – then why not? The pragmatic proof, of course, is whether I have some in my garden shed, and the answer to that question is yes.

The Skeptic, June 2017[7]

Postscript

David Conley, author of that 2003 critique of seaweed extract sellers, published a letter in the following issue of *The Skeptic*.[8] Under the headline 'Seaweed woo?' he repeated his more general concern with companies that promote products with secret or unknown ingredients, based on ambiguous or no trials, heavy on testimonials and, when questioned, turn to conspiracy theories. In his view, the seaweed extract industry is a prime example.

Conley reviewed the more recent papers I cited, concluding that while some 'statistically significant' benefits appear to accrue from the addition of seaweed extract, there is little cost-benefit analysis for its use in the field. He also detects a bias against citing papers critical of seaweed benefits, possibly driven by the researchers being connected to commercial operations.

Conley is suspicious of any product that replaces discounted benefits with new claims – in this case nutrient supplementation with water-holding capacity – and has little time for any company that continues to promote and sell products with dubious value, placing commercial success above all else. He reiterates the decision 30 years before he wrote this letter that seaweed extract fertilisers do not work, full stop, and none of the new evidence convinces him otherwise.

While I am an 'algal guy', my decades of published research on freshwater algae does not include anything remotely related to the use of algae as fertilisers or soil improvement agents. I'm relying on published evidence and the interpretation of other scientists. I agree with Conley's scepticism about a company that simply replaces benefits for commercial gain, but I also accept that if a product works and you don't know why, a good scientist should shift to new hypotheses. So it goes both ways.

The question I ask myself, and the companies and Conley, is which of us is guilty of holding convictions without evidence and being unwilling to change our minds in light of new findings? Maybe all three, to varying degrees. As a supplementary question, if a company is a snake oil merchant, is the product they sell necessarily snake oil? It may be irksome, but the answer must be 'no'.

Synthetic or organic, but minimise harm

It is the ends, not the means, when you garden responsibly

The word 'organic' began its life in the lexicon as a collective term for the chemical compounds extracted from, or existing within, living organisms. With the advent of organic chemistry, the word came to include chemical compounds containing the element carbon, with some caveats which meant that methane was included but not carbon dioxide and diamond – which are considered 'inorganic'. The distinction between organic and inorganic in chemistry remains a little wobbly, but organic is primarily used for carbon compounds associated with living organisms.

Nowadays we more often see the term 'organic' used for food and farming associated with production carried out in a more traditional manner without recourse to synthetic fertilizers and pesticides. This has been described as a 'holistic' approach, harking back, I gather, to a 19th century use of the word 'organic' for a series of coordinated parts forming a whole. That whole is usually said to exceed the sum of its parts, but as an unrepentant materialist, I'm seldom convinced of that.

While the experts at the Food and Agriculture Organization of the United Nations (FAO) acknowledge there are many definitions of what constitutes organic agriculture, they find that all converge on excluding things such as 'synthetic fertilizers and pesticides, veterinary drugs, genetically modified seeds and breeds, preservatives, additives and irradiation'.[9] Soil fertility and pest and disease prevention in organic agriculture must be achieved through other means, typically mechanical or through recycling of once-living things, taking into consideration the specific site and the surrounding environment, in a holistic way, if you must.

The Australian National Standard for Organic and Bio-Dynamic Produce[10] thankfully does not use the word 'holistic' in its definition of organic. Although, it does allow homeopathic preparations – which I suppose would do little harm given they are ineffective and contain mostly water – and excludes genetically modified (GM) plants.

The standards for biodynamic certification are even more questionable, including a requirement to add preparations that 'activate soil and plants, develop soil structure and enhance the nutrient cycles', as well as 'enliven the soil'. While all this is fine in principle, these preparations must be created according to the fundamental principles of the peculiar Rudolf Steiner, as delivered to scientists and farmers at Koberwitz (Kobierzyce), in Poland in 1924. That is, unlike science and more like a religious text, these concoctions cannot be altered or rejected, even in light of new evidence..

All definitions of organic (and biodynamic) farming do not allow the use of 'synthetic' chemicals, defined in the Australian National Standard as 'substances formulated or manufactured by a chemical process or by a process that chemically alters compounds extracted from naturally occurring plant, animal or mineral sources'. This I take to mean a human has deliberately altered something chemically or physically, as opposed to say a worm or yeast. This distinction presumably derives from a moral judgement that chemical reactions or mechanical adjustments taking place in a human setting (a factory?) are intrinsically bad, while those outside or only involving non-humans are good. Possibly a fair rule of thumb but obviously not always true.

The more important question, to my mind, is whether organic plant production is good for the environment and fulfils its intent to do no harm. I have cited David Conley already in relation to my seaweed extract conversion,[11] and while like me he supports a thoughtful and environmentally responsible approach to agriculture, he has no truck with organic farming 'based on anecdotal evidence (at best), cobbled together with science and pseudoscience and mixed with a strong dose of an appealing ideology'.[12]

For arguments sake, let's say the science is good and the ideology peripheral. Should you buy authentically organic food? We already know to question the carbon loading of local food should it be grown in an air-conditioned glasshouse, sometimes giving it a larger environmental footprint than imported food. Surely though, organic food, grown and ripened in the sun nearby, is always better for the planet?

According to a 2016 article in *New Scientist* by Michael Le Pages, the answer is 'no'.[13] While organic farms should support a greater variety plants and animals, because they need more space than conventional farms to produce the same amount of produce, they displace otherwise non-farming habitat (such as clearing rainforests in the tropics). Greenhouse gas emissions can also be higher in organic farms because animals take longer to raise and need more land, *and* because they don't take advantage of genetically modified (GM) plants that may be more efficient to grow.

Andrew Masterson summed all this up in a feature story for *The Age* at the time, as 'there's an emerging consensus among scientists that if we're ever going to feed the world's burgeoning population while simultaneously keeping large bits of it green and wild, we won't be using organic techniques to do so'.[14] More recently, *Guardian* journalist and provocateur George Monbiot cited a recent study showing that 'if England and Wales became entirely organic, [its] land footprint would grow by 40 per cent'.[15] He added later:

> I have come to see land use as the most important of all environmental questions. I now believe it is the issue that makes the greatest difference to whether terrestrial

> ecosystems and Earth systems survive or perish. The more land we require, the less is available for other species and the habitats they need, and for sustaining the planetary equilibrium states on which our lives might depend.

There are counter-arguments of course. In a response to the *New Scientist* article, Tom McMillan, Director of Innovation at the UK Soil Association, argues that it is not just about the efficiency of producing food but also the need to eat different kinds of food and, in particular, less, or better, meat (again, there are heated arguments about whether any meat production is good meat production on both environmental and welfare grounds). Making less food waste would also help, says McMillan, and he disputes the advantages of GM crops. In his view, the increase in wildlife on organic farms is important and more than offsets the lower yields (in a ratio of 50:20). Moreover, in the long term, benefits to the environment from organic farming will continue to accrue, unlike the alternatives.

Clearly, depending entirely on organic farming to produce enough food for the world's growing population of humans would be nearly impossible, even if better for the environment locally and contributing to better overall world health. The same land use arguments can be made for home gardeners, although unless you are growing commercial quantities of vegetable and fruit, you are unlikely to purchase and clear land that might have greater environmental values. Assuming you are already gardening, the size of the plot shouldn't change if you use organic rather than other techniques. Doing less 'harm' to the environment will generally be a good thing, although you do need to assess the cost-benefit of any organic addition or omission.

Adding compost and the addition of shredded plant material has many advantages but it may also introduce disease and weeds. If snails eat all your plants, or poor garden technique leads to low yields, you will have not only wasted water (and other resources) on your own unsuccessful crop but exploited the additional water (etc.) required for the vegetables you buy to replace them. That's a net drain on the planet. I've been guilty of this with tomato plants.

More importantly, just because a chemical comes from a plant without human synthesis doesn't mean it is good – think deadly nightshade, hemlock and oleander for starters. The same is true vice versa – I'm thinking here of a thousands of drugs that have extended our lives over the last few centuries. As in medicine, the risk with 'herbal remedies' is uncertainty over dosage and lack of controlled testing.

A brief overview of organic gardening in 1969 highlighted chemicals in use at that time because they were derived directly from plants but today considered unsafe.[16] The commonly used rotenone, for example, is more toxic than many synthetic insecticides and used outside gardening to kill fish. Nicotine is similarly toxic, and more so than discontinued insecticides such as chlordane and DDT.[17] Thankfully nicotine is no longer recommended for any gardening situation and is excluded explicitly in the latest Australian National Standard[18] but there are plenty of untested chemicals in use.

So-called organic insecticides have had mixed success. Apparently, 'dormant oil' will control a few, minor pests on fruit trees.[19] Boric acid and 'diatomaceous earth' work fine indoors but are of little benefit in the garden where they are subject to rain, wind and higher humidity. Introducing biological controls such as ladybugs and various wasps works better in more 'natural' ecosystems where they can become established with their food and hosts.

In the case of replacing synthetic fertiliser with manure or other organic compounds, quantity and toxicity may not matter as much. Still, misapplication can still result in a shortage or surplus of a particular nutrient, or an unexpected chemical change in the soil. I've made mistakes myself with mulch, changing the acidity so dramatically it killed some of my plants almost overnight.

Synthetic pesticides and herbicides are usually formulated and tested to control insect or disease outbreaks without causing harm to other animals and plants. In the best cases, they leave little residue in the soil and require minimal application. I know incentives are there for unscrupulous companies to exploit this opportunity and not to care about environmental consequences. We should fight against that, but not 'throw out the baby with the bathwater'.

I prefer a more nuanced approach. A big 'yes' to minimising harm in our neighbourhood and across the planet, but 'no' to *prescribing* processes and materials or excluding improvements through techniques like genetic modification. Assess each on a case-by-case basis. Harm minimisation may be best achieved through selective use of safe synthetic chemicals alongside other more 'natural' techniques.

I know this approach flies in the face of those who call for tighter controls on what we label as organic, but that in the end is a hollow, intellectual pursuit. Harm, like happiness, is not an absolute term and as we see in palliative medicine, 'do no harm' can be a difficult concept to apply. Whether you go safe synthetic, organic or a bit of both, minimise your environmental footprint at home and abroad.

To return to one of my familiar themes, whatever route you take, learn to live with blemishes and a few plant losses.

April 2024

Uncompanionable planting

A good companion can attract or repel the undesirable, or harbour their nemesis

Companion planting is when you grow one plant species close to another, with the intent that at least one of them will benefit from this arrangement. Familiar couplings include tomatoes and basil, pumpkin and corn, beans and radish (or beans and cabbage), beetroot and broccoli, and various combinations to ward off cabbage white butterflies from their hosts.

Companion planting is not crop rotation. This is when you plant heavy feeders such as tomatoes and cabbage in the year following a crop of nitrogen fixers, such as legumes. Companion plants are grown together, at the same time, in the same place.

This means they are competing for the same resources, for sunlight, water and nutrients. So the benefits need to outweigh the costs. In a perfect liaison, both species will thrive. You might even be able to harvest and enjoy the two plants in the same meal, as in the popular tomato and basil combo.

Typically, though, a companion plant is introduced to help reduce insect or other pest attack in a favoured crop. The main reason for planting basil near tomatoes is to benefit from basil's chemical unattractiveness to things like tomato hornworms, whiteflies and aphids, all bugs you don't want too many of on your tomato plants. While this seems to work, even advocates admit that you need a lot of basil for each tomato plant, not just a plant here or there.

Another commonly cited companion for tomatoes is marigold, to reduce root-knot nematodes. The roots of marigold have been shown to release chemicals that suppress some nematodes, although they do encourage others. The problem here is the destination of irritated nematodes. If you plant the two species together, they will most probably seek out the more pleasant host, your tomato plant. Better to treat these two species as rotating crops, in separate years.

Sadly, cabbage butterflies (and their caterpillars) are difficult to shift through companion planting. A much-cited study by Virginia

State University, from 1978, showed that *no* co-planting of cabbages with marigolds, nasturtium, pennyroyal, peppermint, sage or thyme – all said to repel leaf-eating insects – would reduce insect damage to cabbage leaves.[20] In some cases, there was more.

In a rather cute study of female cabbage root flies, researchers from the United Kingdom watched them hop (or spiral) and walk (or run) across the leaves of interplanted cabbage and clover.[21] It turns out that in mixed plantings, the flies visit both species, but an individual landing on a clover spends more time there and doesn't bother seeking out the cabbage. Contrary to what was assumed before this study, the flies appear to be – as the authors put it – *arrested* rather than *repelled* by the clover. The companion in this case works as a decoy or sacrificial host.

A companion plant doesn't necessarily have to attract or repel more than its buddy. Simply by bulking up the number of hosts you may dilute the impact of an unwanted insect visitor on your crop. Of course, this may mean you end up with more of the insect and just as much damage, but in some cases, it seems, it pays to be one of many.

A final mode of action for a companion plant – aside from repelling and attracting nasty bugs – is to encourage *predators* of some kind (other insects, or perhaps birds) to feed on the species eating your favoured plant. There is some evidence for borage, mint and rapeseed acting as attractors for predatory insects. As with basil and tomatoes, success in your home garden will depend on the relative numbers of plants and their arrangement, and the complex ecosystem you can create of 'good' and 'bad' bugs.

In practice we seldom know how many of the companions are needed and in a particular setting whether their benefits outweigh competition for resources. Companion planting may be too complex to be practical. Rather than seek companions to help fight off pests, select plants that simply coexist but don't demand too much of any limited resource (e.g. water). Ideally, they attract pollinators and add colour and beauty to our garden landscape. In the best of all worlds, they do all this and are edible themselves.

Maybe 'intercropping' – the growing of two or more plants together – is a better term. It is used in some modern agriculture to make best use of all available environmental variables and inputs and to increase total yield. For example, simultaneously planting of cowpea and scarlet eggplant in Ghana results in a similar yield of scarlet eggplant (to when it is planted alone) but with additional cowpea to use for food or fodder.[22]

Species diversity in nature is generally a good thing (although see my comments on this elsewhere in this book)[23] and judicious interplanting of anything with lots of flowers will most likely help your crop. The more distant the species the better – you don't, for example, want to plant tomatoes and potatoes together because being in the same plant family means they are more likely to succumb to the same pests and diseases.

Aim too for a little clutter, in order to attract as many different insects as possible: potential pollinators as well as predatory bugs. And then, may the bugs and plants you want thrive!

April 2024; Gardening Australia, November 2019[24]

Over your dead body

(**Note:** please be advised this article mentions human placenta.)

> *Perhaps the best companion for a plant is an animal. Preferably dead.*

There are plenty of important decisions to make immediately after giving birth: breast or bottle, disposable or cloth (nappies), and, for some, which school. One of the most urgent, though, is what to do with the placenta – that part of the uterus expelled after the child is born but up until then a conduit for nutrients and oxygen to the foetus for 9 months.

I remember being told by a friend at the time of our first born that we should eat the placenta or plant it. Placentophagy – eating the

placenta – carries some risks through transmitting unwanted bacteria and viruses and there are no proven health benefits to doing so. Besides it didn't appeal to me.

Burying is a ritual in many cultures, sometimes associated with prayers, the burying of other significant items, or the planting of a tree. I don't have a need for prayers and can't see much value in storing my valuables underground, but anything that creates an opportunity for a new tree is a good thing.

Even better if the placenta will help the tree thrive. Basically, it is a piece of meat, and we happily fertilise gardens with blood and bone. Carnivorous plants, such as sundews, Venus flytraps and pitcher plants, digest insects and other animal visitors to extract nutrients from their carcass. There is clearly something good to be had from animal flesh.

In nature, in and outside our gardens, it happens all the time. Every time a mammal, bird, reptile or fish dies, let alone billions of insects and smaller critters, their remains will decompose to the benefit of us all. This is recycling and our living Earth depends on it.

At home, we are a little more squeamish about such things. I should at this point remind you that in most jurisdictions, it is illegal to bury a human body in a place other than a cemetery; in Victoria, you can spend up to 5 years in prison for committing this offence. However, I gather you are free to bury your pet dog or cat, or perhaps turtle or hamster.

Yet we tend not to gather dead animals for composting in the way we might collect leaves and debris. That has a lot to do with good health and foul odours, of course, but if done carefully the occasional (non-human) buried body will benefit your garden.

You'll find some rather gruesome recommendations for composting dead animals[25] – such as dismembering the corpse and puncturing its belly – but even with very little preparation, a dead animal-based compost can be useable within 4 to 6 months (although you might wait a year to be sure).

Alternatively, you may want to simply bury your dead animal directly into the garden. This is, after all, what happens in 'nature'. Overseas researchers who study such things find that a season or two after the death of large animal, the amount of plant material growing on or near a corpse can be up to five times higher than surrounding areas, and nutrients such as nitrogen and phosphorus up to eight times greater.[26] There are also likely to be more insects both to eat plants and eat other insects that eat plants (although some studies show numbers of plant-eating animals are reduced near carcasses). Those high nutrient levels can remain for up to 10 years.

Bringing it closer to home, a study on kangaroo carcasses showed that the changes in soil chemistry and therefore plant diversity above the ground near the dead carcass remained for up to 5 years.[27] Unfortunately, that benefit was mostly to weedy annuals, albeit ones already growing nearby. In your home garden, you can make sure only your favourite plants thrive, so it seems a few decaying mammals would generally be a good thing.

As to our own post-birth decisions: breast, cloth and thriving in the local schools (at least at first). We did plant a tree but in the euphoria of welcoming new life we left the placenta at the hospital. To be honest, that was always going to be our decision.

April 2024; Gardening Australia, November 2019[28]

Glyphosate, punished for the sins of its inventors?

Roundup, the herbicide that can be used for good and evil

As I began writing this essay, a case was heard in the Federal Court in Melbourne as to whether glyphosate – the active ingredient in widely used weed-killer Roundup – is a carcinogen.[29] It was a class action against Monsanto, now part of Bayer, brought by Kelvin McNickle on behalf of 800 Australians with cancer of the lymphatic system (non-Hodgkin lymphoma). McNickle contracted non-Hodgkin lymphoma at the age of 35, after 20 years exposure to glyphosate.

As things stand, glyphosate – or N-phosphonomethyl-glycine – is not considered a significant carcinogenic risk to humans in any country, including Australia (although there are restrictions on its use in some jurisdictions). This is despite the World Health Organization's (WHO) International Agency for Research on Cancer listing it in 2015 as 'probably carcinogenic to humans'.

That WHO decision was controversial due to perceived conflicts of interest and interference from interested parties,[30] and for the wording of its judgment. Glyphosate was assessed as a 'category 2A compound', meaning 'there is strong evidence that it can cause cancer in humans, but at present it is not conclusive'. Other substances in this category include eating red meat, drinking hot beverages and being a hairdresser.[31]

Concern about the potential impacts of glyphosate on both human health and the environment more generally began soon after its discovery by a Monsanto chemist in 1970. The debate was invigorated by the WHO decision in 2015, particularly around how much exposure to glyphosate would be considered acceptable, even when wearing full protective equipment. McNickle argued at trial that he and his kit were 'saturated' in Roundup at the end of each day.

Glyphosate kills plants by inhibiting an enzyme required to produce proteins needed for healthy and sustained growth. Treated plants become stunted, lose their green colour and often wrinkle or become malformed, before dying within 4 to 20 days. These metabolic pathways are only found in plants and some microorganisms but not in mammals.

According to regulatory organisations, glyphosate in dead leaves and soil is quickly broken down by bacteria, although it may persist for up to 6 months.[32] During that time, it binds closely to soil particles and is therefore unlikely to move into groundwater.

Outside the court rooms and hospital wards, gardeners have mixed feelings about glyphosate. I recall a post on social media a few years ago by bush regenerator Peter Dixon, incensed at the false information, pseudoscience and dodgy sources feeding into an inherent hatred of Monsanto. The 2017 social media post ended with praise for 'effective herbicides', noting that both agriculture and gardens benefited from their responsible use.

To be fair, Monsanto deserve our condemnation for some of their historical practices and the less than responsible way they marketed and controlled what they called 'Roundup Ready' crops.[33] Genetically modified plants of soy, maize, canola, sugar beet, cotton and alfalfa were bred resistant to glyphosate, so the weedkiller could be sprayed onto crops, killing only the weeds. This led to questionable business priorities and practices in the quest for profit.

Responses to Dixon's social media post included references to toxicity on human embryos and bees, a general call to end our 'addiction' to pesticides and herbicides, and acknowledgement that even if the product was safe, it was hard to forgive Monsanto. I should also note here that glyphosate is prohibited for household use in number of European countries and banned entirely in Vietnam and some parts of the United States.

One respondent mentioned other potentially toxic ingredients in Roundup, such as the surfactants that help it adhere to plants. The regulator's fact sheets make mention of two surfactants that may be toxic is digested by animals, such as frogs. A 'bioactive' variant of Roundup was released to address this particular problem. Dixon himself favours less use of herbicides but observes that 'weeds cause more ecological harm that glyphosate'.

There may be reasons other than toxicity for avoiding glyphosate. For example, overuse of Roundup has driven the evolution of glyphosate-resistant weeds, up to 35 of them in 2018.[34] With reference to my essays in Chapter 2 on the often unwarranted demonisation of weeds, you might push back on any compound that kills plants indiscriminately (apart from specially bred exceptions). For that matter, killing indigenous and exotic plants indiscriminately.

I want to be clear here that I am not encouraging the use of any herbicide or pesticide without due care and responsibility. A recent study showed that despite not being volatile and therefore airborne like many other pesticides, glyphosate does indeed accumulate in people (in this case pregnant women) living nearby to treated crops.[35] And the impact of chemicals such as glyphosate on pollinators and other insect dependencies (for plants and people) is a whole separate question.

Like the introduction into farming of genetically modified organisms (the topic of my next essay), decisions on glyphosate use should be made based on facts, not vendettas or philosophical leanings.

April 2024

Postscript

In July 2024, the class action against Monsanto mentioned at the beginning of this essay was dismissed by Justice Michael Lee, who found there was no evidence provided to the court establishing a link between the use of glyphosate and non-Hodgkin lymphoma.[36]

Commodified genes or just more of the same?

Genetically engineering and conventional plant breeding both manipulate genes to produce more productive crops

In 2007, when I was a member of the Australian Academy of Science's National Committee for Plant and Animal Science, we prepared a position statement for the Academy on the use of gene technologies to modify plants; that is, the techniques used to create what are called genetically modified organisms, or GMOs.[37]

GMOs is an odd name, given nearly every crop we grow and eat has had its genes modified by humans in some way. All genes have been *modified* by evolution. Still, we all know it refers to more recent advances in genetic manipulation, making it possible to change the plants in a more precise and far-reaching way than ever before.

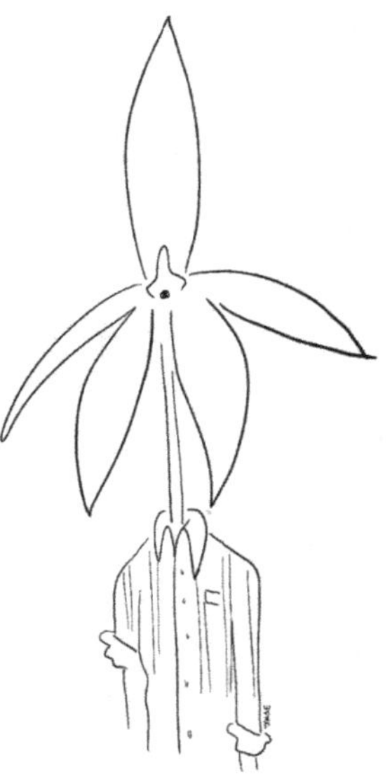

The statement – written largely by our Committee chair, academic and leading agricultural scientist, Professor TJ Higgins – supported 'the responsible and ethical' use of gene technologies in agriculture. We saw the benefits of GMOs as including the not insignificant need to address malnutrition, improve environmental sustainability and to produce greater crop yields on less land.

While uncertainty around the use of GMOs was acknowledged, as must surely be the case with any technology, if those risks can be addressed, we considered that there were too many benefits for the technology to be dismissed. The introduction of GM cotton into Australian agriculture had already reduced pesticide use, and over coming years, GMOs were sure to be a source of new medicines, help with the remediation of polluted soils and waterways, and assist in the control of pests and weeds.

The Committee found no evidence of ill health in humans from the consumption of GM foods and considered the regulatory system in Australia up to the task of making this risk unlikely. That said, we supported labelling of GM-derived foods to help 'consumers making deliberate dietary choices', as long as the labelling was based on good, evidence-based science.

Regarding environmental impacts, the committee acknowledged that all crops and pasture plants have the potential to harm neighbouring ecosystems, with or without GMs. Scientific evaluation and adequate trials are critical before any release and must continue. The Committee added a call for more attention to business ethics and for continuing 'public scrutiny and safety of genetic research', including strong regulations and regulators.

In hindsight, the statement lacked a little finesse and clarity of language, but the intent remains true today, as I was reminded recently, when I read Mark Lynas' *Seeds of Science*.[38] Lynas, a once radical activist against the introduction of GMOs, became converted to the cause for some of the reasons I've outlined above.

For Lynas, much of the campaign against GMOs had become ideological and contrary to the facts. The disastrous effects on human health and the environment foreshadowed by the current King of England, Charles III, in 1998 had not materialised from a practice that, according to the then Prince, took 'mankind into realms that belong to God and God alone'.[39]

Lynas re-examined his own views after receiving criticism of an article he wrote in opposition to GMOs, using terms like 'genetic pollution', 'can never be recalled', 'GM superweeds ... run rampant and breed' and 'infest' other crops. One of the rejoinders accused him of creating an 'evil aura' around things that are simply what they are and do what they do, and which are usually benign.

When Lynas read the published evidence, he was chastened to find 'there was nothing much to support [his] claim that there had been "countless cases" of GM crops infesting other fields or otherwise spreading damaging "genetic pollution"'. And that there was 'zero evidence that any genetically modified foods in existence today pose a health risk to anyone'.

In fact, to reduce some of the negative impacts of agriculture on the environment and to improve human health, Lynas argues we should use the best techniques at hand to breed more productive and safer plants. That is, genetic engineering or modification.

Companies like Monsanto have faults – inadvertent or deliberate – but they have also sought to reduce the use of insecticides. Consistent, as Lynas argues, with the view of Rachel Carson in *Silent Spring* who advocated for 'spray[ing] as little as you possibly can'.

As an aside, Lynas feels Monsanto has been unfairly singled out as 'the most evil corporation in the world' given the stiff competition the company faces for this title; particularly given 'its move into genetic engineering was probably the most environmentally friendly thing that Monsanto ever did'.

As mentioned in the previous essay, Monsanto's motivations are open to question, with one of its first genetically modified crops (Roundup Ready soy), a plant able to survive glyphosate spraying and thus supporting its chemical manufacturing arm.[40] This was. at best, a major PR disaster for Monsanto. As Lynas says, 'genetic engineering could have been associated in the public mind from the outset with the reduction of chemical pesticides and might therefore have faced less widespread opposition'.

There were also a few own goals. So-called terminator technology, where a plant is bred to intentionally produce sterile seed, was mooted and tested by Monsanto but did not eventuate. This would have forced consumers to buy new seed each year. This is, however, something that companies like Monsanto pursue in any case to protect their plant patents and rights and, it has to be said, as Lynas does, hybrid seeds that don't breed true in subsequent generations have long been part of gardening and agriculture.

Other criticisms are simply incorrect, but I'll leave you to follow up on them in *Seeds of Science*. Lynas aligns with Oxfam America, at the time of writing his book, which does not espouse a view for or against GM technology, considering each case on its own merits. That means waying up the human rights and fundamental needs of all parties, including farmers and consumers, and assessing the risks of any new technology rationally and responsibly.

The uptake and application of GM crops over the last few decades has been constrained by vocal and provocative protests, and therefore wary governments and corporations. Even so, adoption of the GM technology up to 2014 had already reduced chemical pesticide use by 37 per cent, increased crop yields by 22 per cent and pushed up farmer profits (mostly in developing countries) by 68 per cent, according to data cited by Lynas.

GM crops have also contributed to reductions in atmospheric carbon and an increased in insect diversity. While there are documented cases of accelerated weed evolution due to higher use of glyphosate to control weeds in glyphosate-resistant crops, there is little sign of GMOs moving from crops to locations other than neighbouring roadsides and irrigation channels.

As the Royal Society of London put it in 2016, 'genetically engineering and conventional plant breeding both produce crops with improved characteristics by changing their genetic makeup'.[41] The bottom line is that any risks in growing and consuming GM crops are about the same as those for plants taken directly from nature or modified through more traditional means. This what the Australian Academy of Science stated in 2007, it's what Mark Lynas wrote in 2016, and it remains true today. While we must remain vigilant with every new plant introduction, it behoves us to consider each on its merits.

April 2024

Wood-wide web unravels

'A mat of long, thin filaments that connect an estimated 90 per cent of land plants'[42]

Welcome to the wood-wide web, a place where trees trade, share and befriend others of their species, and perhaps other kinds, through an underground network of cooperating fungal threads. In some extreme renditions, a benevolent social network where plants support one another through acts of kindness and self-sacrifice.

A counter view is emerging among those who should know – the mycologists – that while the wood-wide web is a catchy slogan, it is also an overhyped and overextended metaphor, perhaps 'a fantasy beneath our feet'.[43]

While nearly all land plants have fungi associated with their roots – called mycorrhizae – there is little evidence yet that these

symbiotic associations do more than provide nutrients to a plant in exchange for sugars to the fungus. As to this fungal–plant relationship creating an incipient social network of some kind, that is at best wishful thinking. For a start, even though 90 per cent of land plants may have fungi associated with their roots, that does not mean these filaments *connect* even one plant with another.

In 2023, I attended an online talk by Dr Camille Turong, a research scientist and mycorrhizae expert from Royal Botanic Gardens Victoria,[44] who along with Canadian ecologist, Justine Karst and colleagues, is concerned that the hoopla for the wood-wide web far exceeds the hard science.

Drawing on her early life in Switzerland, Turong frames her response around a saying from her childhood, 'you have put the church back in the middle of the village', which I take to mean we need to de-escalate the matter, returning it to what we know. Or as I might have put it, quoting Wittgenstein from 1921, 'whereof we cannot speak, thereof we must be silent'.[45] This is pertinent to what Camille is saying in her lecture. We like to think trees can communicate, she says, particularly with us.

What we do know is that one plant can associate with more than 30 fungi in mutually beneficial relationships. These mycorrhizae allow plants to live in environments where they otherwise could not or could hardly do so; in arid and nutrient-poor terrain, through drought, and after fires.

Turong points out that these fungi are obliged to be symbionts, while the plants are not. Plant can live without the fungi but perhaps not so well and not in so many places. Because plants typically produce excess carbon in the form of sugars that fungi need to survive, it is a low cost to the plant. In a high nutrient environment – such as in potting mix – fungal partners are not needed.

Here we get a side bar from Turong around the use of what is called 'compost tea', a concoction added to garden plants to help them perform better by encouraging beneficial fungi. The tea contains root balls (with fungi) but also manure, leaf litter, dust and wood chips. Turong's research with other colleagues showed that there were less mycorrhizae in these instances because the plants get enough nutrients from the rest of the brew. She adds that in nature, things are often different, so when nutrient levels are low, as they often are, ectomycorrhizal partnerships are common. The point is that the compost tea has been sold with deceptive marketing.

Which brings us back to the wood-wide web and the even more extraordinary claims made about it. First, are mycorrhizal networks even common in forests and natural settings? The epigraph at the start of this essay implies so, as does the rhetoric about them providing a universal means of communication between plants.

Scientist can map fungal networks through a forest using molecular (DNA) analysis of soil samples, but it's a demanding task. In fact, so demanding that at the time of Turong's talk, only five thorough studies had been completed, in two countries (Canada and Japan), across two forest types dominated by conifers, and involving three fungal species.

Even where mapped fungal hyphae reveal the same individual fungus on adjacent trees, it couldn't be confirmed that there is or was a connection. Networks frequently disconnect across their range when mycelia die, break or are grazed. For Australia, the data has little relevance given that the species and systems studied are all from forests in the Northern Hemisphere. Turong's current view – a hypothesis to be tested – is that long-term persistence of common networks is unlikely in this country.

For the sake of argument, let's say these fungal connections do exist, for some species, in some places. One of the appealing traits of the wood-wide web is its ability of the mycelial network to act as a conduit for a plant to direct resources to a neighbour of the same species, perhaps its own offspring. An even more altruistic sharing has been suggested by some, from species to another.

It is true that in some of the systems tested – spruce trees in the Northern Hemisphere, for example – carbon from one tree is found in a neighbouring tree. Not in Australia so far, but it does happen overseas, albeit in very small amounts. While that carbon could travel through the fungal pipes, it could also get from one plant to another through root exudates diffusing through the soil or even close contact between the roots.

In one of the most cited papers on this phenomenon the authors canvas all these options, and can't confirm or reject the fungal explanation.[46] Another experiment using potted plants showed that similar rates of carbon transfer occur even in the absence of fungi. Studies on seedlings have given variable and inconclusive results.

As to the 'mother tree' idea, where a revered old sentinel sends messages or sugars to its offspring, there is no definitive evidence of this occurring. Dying or stressed trees do release nutrients, which neighbouring trees may use, but that doesn't imply any mode of action requiring fungi. Evergreen trees can also release or lose sugars in winter which may be taken up by nearby deciduous trees (without their own source of food during winter). Again, this may be simply a benefit of proximity, not willing fungal partners.

Another perspective on this carbon transfer, if and when it does happen, is that it is driven by the needs of the fungus to move the carbon around its own mycelium, to extend its own growth and reach – not to benefit plants. As fungal advocate Merlin Sheldrake has put it, fungi are more likely to be farming their hosts than the reverse.[47] For now, all we know is that to a limited extent, carbon moves around underground.

Finally, what about this signalling and even 'talking' across these mycelial networks? There is some evidence of fungal threads being the medium for distress or warning signals between adjacent bean plants. There are also tantalising hints of this in potted plants – in this case pines, from the Northern Hemisphere – but with the caveat that when roots touch, this communication stops. Again, there are no peer-reviewed field studies and only a limited range of species have been tested.

Overall, Truong echoes a sentiment expressed by her colleagues: less hype and more hyphae. When I asked whether she may be guilty of reverse bias, Camille says that she began these studies excited by the prospect of discovering the extent and importance of the wood-wide web. It was with some disappointment I think that she finds herself part of a movement to deflate the whole concept.

Like all of those fighting for this reset of the wood-wide web, Camille wants fungi to be better recognised and appreciated. Life in the underground world is often overlooked, and the wood-wide web and similar concepts are at least drawing attention to a much neglected and underappreciated part of nature.

After researching for this piece, I'm a little shy of making any claims about mycorrhizal fungi. But even if the numbers are inflated, fungi clearly have an important role to play in our planet's health. By one estimate, the 450 quadrillion kilometres of fungi in the Earth's top 10 centimetres are responsible for sequestering 5 billion tonnes of carbon dioxide, or 75 per cent of terrestrial carbon dioxide on Earth.[48] Annually, according to another source, this translates to 36 per cent of global fossil fuel emissions being held (at least for some time) within mycorrhizal fungi.[49]

Camille and other critics have no problem with the wood-wide web remaining as a rallying cry for fungi but they plead for more constraint. Remove the bias, they beg, and pursue more experiments and evidence gathering, more critical analysis of all alternatives. A familiar cry by scientists, but here part of a much-needed reset.

April 2024

Rhizosphere revealed

The subterranean world around plant roots does better when we butt out

It is said that a gram of soil, just a few crumbs, can contain more than 1,500 kilometres of DNA,[50] this being the genetic instruction strip found in all living things, from bacteria to bison. The ground below us is teeming with life.

Obviously, there are few bison in a gram of soil but there can be up to 200 metres of branching fungal threads, a billion or so bacterial cells, and an equally beguiling assortment of protozoa, algae and other microorganisms. Gather up a cupped handful of soil, and you may be holding as many microorganisms as there are people in the world.

I'll let you take that in for a moment. So much life. Because we can't see, feel or smell it, we barely acknowledge its existence.

While many of those subterranean microorganisms are fungi – part of that (in-)[51] famous wood-wide web – mycelial networks are not the only show underground, even within what we call the root biome or rhizosphere.

Plant roots have numerous obligate biological partnerships. Well known to agricultural students are the nodules on legume roots,

containing the nitrogen-fixing bacterium called *Rhizobium.* These bacteria convert atmospheric nitrogen into oxides, a form plants can access readily.

Feathery clusters of fine roots, often at the surface of the soil, help plants living in nutrient-poor soils access phosphate. These proteoid roots – commonly found in, but not exclusive to, the protea family – also attract a rich assemblage of bacteria to assist.

Ardent environmentalist and journalist George Monbiot describes the rhizosphere more generally – the chemicals and life forms living on and around roots – as a tree's 'external gut'.[52] Up to 40 per cent of sugars produced by a plant through photosynthesis are released into the rhizosphere to support this outsourced digestive system.

Appropriately then, just like our human gut, there are a lots of bacteria. Along with plenty of things with legs, if I can put it that way, such as insects and spiders, but also amoebae, slime moulds, and worms of various kinds such a nematodes. But bacteria are by far the most diverse component of this underground universe.

From DNA sequencing of the organisms living around the roots of thale cress (*Arabidopsis thaliana*), the lab rat of the plant world, we know there are representatives of at least six bacterial 'phyla'.[53] Phyla – or when there is one of them, a phylum – are a taxonomic category at the level of the flowering plants, that is, all the floral variety you find in your garden – from geranium to gum tree to grass – belong to the flowering plant phylum, Magnoliophyta.

The Magnoliophyta in turn rolls up into a higher group with other so-called vascular plants (ferns and conifers), then with all 'green plants' (with mosses and some algae), then with animals, fungi, other algal and various 'protists' to eventually form the Eukaryota. This is where you and I sit as well, the group containing pretty much everything made from cells with a nucleus and 'membrane-bound organelles'.

Bacterial cells lack a discrete nucleus and organelles so are classified outside the Eukaryota. They contain as much genetic and evolutionary diversity in each of their phyla as the flowering plants. So there are not only billions of individuals in that gram of soil but far more variety than among the plants in a garden or forest.

Most of these bacteria live by breaking down plant sugars, although some are partly or fully photosynthetic (i.e. able to produce their own sugars from sunlight and carbon dioxide). In the latter category are the Cyanobacteria, sometimes also called blue-green alga, my excuse for including this essay in a book of botanical scepticism.[54]

As to whether the flora and fauna of this 'external gut' is much different in soil without roots... perhaps. The phyla represented are probably the same as in any gram, or cup, of soil, but then there is likely to be dead or alive plant material in most soil so it's hard to confirm this. Some of the bacteria found near roots also colonise wooden sticks, implying that lignin in roots may be the attractive thing rather than some kind of trade like the mycorrhizal fungi have based on plant exudates.

It is still unclear if those arriving from surrounding areas are there by chance or somehow actively recruited by the plant – or, to take the reverse perspective, active colonisation by the microbe. Although most seem to arrive from soil nearby, 'some are bequeathed from previous generations via the seed'.[55]

What then can you do, or not do, to care for this rich assortment of microorganisms beneath your feet?[56] They say if you rip a plant out of the ground and look at its roots, a healthy rhizosphere will be evident through plenty of soil adhering to the roots. This material adhering to the roots is called the rhizosheath.

I would suggest the extent of the rhizosheath depends as much on the root system of the plant you pluck from the ground – for example, the extent of fibrous and fine roots, degree of root hair growth and the age of the plant – and the kind of soil you pluck it from – sand is unlikely to adhere no matter how much exudate and how many microbes inhabit the rhizosphere. Also, how wet the ground is at the time of extraction. In any case, I do not recommend you uproot your favourite plant to check.

There are plenty of products sold as 'biostimulants', said to provide and support the microorganisms you need around your plant roots. Because most are already found in soils I wouldn't bother trying to add more, and as to encouraging them through chemical additions, I'd leave them to the plant to dispense, as required.

Plants will adjust the amount of root exudate depending on what they can afford or what they need at any particular time of year. If you must add something, a regular application of compost will help maintain good soil structure, hold water and is unlikely to do much harm to the rhizosphere (or rhizosheath). Adding herbicides and pesticides almost certainly will do harm if applied in excess or poorly targeted in application. Similarly, tilling is more likely to upset than help the rhizosphere.

The best advice is to pull up a chair and enjoy what you see above the ground, knowing that the world beneath toils away successfully without your help.

April 2024

The Moon factor

Lots of gardeners swear by Moon planting, but does the science back it up?

With NASA preparing for humans to return to the Moon's surface in 2025, I feel brave enough to revisit the topic of celestial horticulture. A few years ago, I argued that planting by the cycle of the Moon has no direct bearing on the growth of a plant here on Earth, yet I continue to be asked about the veracity of night-flowering cacti blooming only on the full moon and other potentially Moon-induced plant behaviours.

It seems that the Moon as a gardening talisman just won't go away. Do you plant your garlic or potatoes when the Moon is on the wane? Or prune your fruit trees with a waxing Moon? Perhaps you add fertiliser during the new Moon but transplant when it's full? You are not alone.

This is despite the stark reality that there is no reliable evidence for any relationship between the lunar cycle and plant 'behaviour'. Added

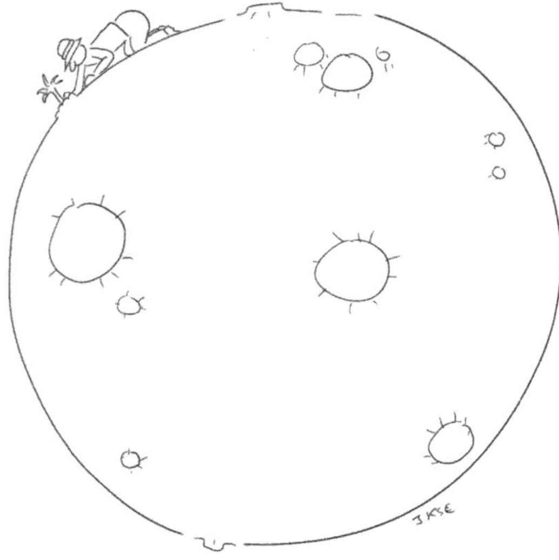

to that, there is no physical reason why the Moon *should* influence plants. I cannot discount the possibility, but as things stand today, here are five lunar truths.

The gravitational pull of the Moon at the Earth's surface is tiny

It has been claimed that the gravitational pull of the full Moon will draw sap from the roots to the leaves of a plant. This is sometimes cited as the reason why you should plant 'above-ground' crops when there is more moonlight (between the new and full Moon – a waxing Moon) and 'below-ground' crops when there is less (between the full Moon and the next new Moon – a waning Moon).

It is tempting to assume a full Moon will pull sap up a tree in the way it pulls the Earth's oceans to higher tides. If the Moon is powerful enough to create tides in the Earth's oceans and large lakes, then surely this is not only possible but likely?

In actual fact, the Moon's gravitational force is minuscule, and it only has consequences (tides) on something as large as the oceans – which, when you consider that they are all interconnected, form a very large object indeed.

Other, still huge, water expanses, such as the Mediterranean Sea, show negligible tidal movement. This puts a tomato plant, or even a tall tree, in perspective. To bury this idea further, the gravitational pull of the Moon at the Earth's surface is a thousand times less than the pull of a 1 kilogram block of butter. That is, it is negligible, in the extreme.

If the Moon's gravity *were* important, you'd also expect two sap rises and falls each day to match the daily tides, as well as the extra bump with a spring tide. That is not something scientists observe.

Moonlight is not the same as sunlight

First, sunlight is, on an *average* day, 130,000 to 400,000 times stronger (light intensity, measured in lumens or lux) than the light reflected from a full Moon. Even on a completely cloudy day, sunlight is many thousand times stronger than moonlight.

Secondly, experiments have shown that 'short-day' species – ones that flower or respond in some way to long periods of darkness – are not influenced by a full Moon. In other words, light from the Moon does not break the dark/night cycle. In plants, it seems, the photoreceptors are not sensitive to this very weak and occasional moonlight.

Corals behave differently to plants

There *are* some living things on Earth that are synchronised with the lunar cycle.

Corals spawn synchronously a few days after a full Moon so that they overwhelm their predators and maximise survival of eggs, sperm and, ultimately, new corals.

Spawning only occurs when the water is warm enough, and is then triggered by what seems to be a mix of day length, tidal height, salinity levels and, indirectly, the lunar cycle. Coral, and some coral fish, have a 'photoperiodic' response, like when plants initiate bud formation or leaf drop with shorter or longer days. The corals appear to detect the increasing length of darkness between sunset and moonrise, triggering spawning a few days after a full Moon.

Given that plants have chemicals called photoreceptors to detect and respond to day length, it's not implausible that they could be responding to moonlight. However, again, for now, we have no evidence of a 'moonlight receptor'.

Magnetism and polarisation of moonlight are unimportant

Red light is known to trigger photoperiodic responses in plants, and some researchers have suggested that polarised red light might influence a plant's behaviour. Polarisation of moonlight in the red spectrum does occur as it travels to Earth, but to a very small degree, and there is no evidence that it is registered or responded to by plants.

As for magnetism, unlike the Earth, the Moon has no magnetic field. Even if it did, its influence on Earth would be as insignificant as the Moon's light or gravity.

Our memories are selective

Given all this, why do we think the Moon affects plant growth? A 'fullish' moon will persist for about three nights of the Moon's roughly 29-day cycle, meaning that there is about a one in nine chance of a night-flowering plant doing its thing 'on a full Moon'.

Couple that with our tendency to remember only positive associations – that is, we forget the times when a flower appeared on a dark night because there was no remarkable celestial event to trigger our memory – and there may be no phenomenon to describe anyway.

This seems to be the case with night-flowering cacti – such as queen of the night, *Epiphyllum oxypetalum* – which are sometimes said to flower 'more often' with a full Moon. Sadly, another case of wishful thinking. Keen cactus growers tell me that if you watch enough of these flowerings, about one in nine track the full Moon, as you'd expect. For the record, mine flowered last on a waning crescent Moon in late February.

Still, having dissed the Moon as a horticultural aid and influencer, I can't think of a good reason *not* to use the lunar cycle to plant if you have the correct season, good soil and adequate water. Maybe the moonlight will improve your mood and thereby the beauty of your garden.

Gardening Australia, July 2022[57]

Postscript

I continue to meet people convinced that planting by the lunar cycle reaps horticultural rewards. The theories and results differ but generally the full Moon features in the planting and the rewards are healthier plants or more bountiful harvests. Historical references and testimonials are often cited.

Indeed recently, my writerly friend and ex-Trustee of Royal Botanic Gardens Kew, Jonathan Drori, sent me a clipping from the learned scientific journal, *Nature*, headed 'Growing seeds by moonlight, and a star shower seen at sea'. This is an item from 1923, republished from the journal's archive, describing an experiment into the effect of moonlight on the germination of mustard seeds. The investigator found that seeds exposed to the light of the Moon were 15 per cent more likely to germinate. This was attributed this to polarised light, which they say was supported by testing with daylight reflected or through a prism.[58]

Yet all of what I wrote above is still true.

5

Observing and cataloguing nature

As will be abundantly clear by now, I am a taxonomist, which is why I care about what we call things, and why we call them what we do. I like to observe and catalogue nature; to name things, work out how they relate to other named things and then ponder over how they got to where they are today. Applying the correct name is for me a key to what is known, and unknown, about a plant (or alga or fungus). This is not the only key, and it doesn't open all doors, but it's a good place to start.

Threaded through this chapter are essays on plant nomenclature, perhaps my favourite pastime. But while I argue passionately for those whose job it is to document and interpret nature – the botanists and taxonomists – I accept that in many cases, the best name is one you can remember. Also, that the name should be a starting point, not the end game.

In that context, when I wrote my book on Australian seasons, *Sprinter and Sprummer: Australia's Changing Seasons*,[1] I was more interested in readers observing change in the world around them than I was in them adopting my new seasonal system. If we don't notice and acknowledge the local seasonal changes, how will we ever accept the bigger and more sinister changes caused by the climate crisis?

Here, I offer two further perspectives on seasons: first, some reflections on seasons in London written in the early 2010s while I was writing *Sprinter and Sprummer* , and then a piece from 2023, where I reflect on the evolution in my thinking since the publication of that book in 2014. As a sceptical botanist I need to be open to changing my own mind as well as others.

Speaking of change, Victoria's floral emblem – do you know what it is? – is pretty enough in the bush but underwhelming and difficult to display on formal occasions. I have a few suggestions about what might replace it. Also in need of replacing are some of those scientific names I obsess about. I run a sceptical eye over recent attempts to rid plant nomenclature of inappropriate and offensive eponyms, siding in the end with those calling for change. Again, absolutism is not in our best interest if we want names to be used and valued.

As to 'eradicating extinction', how is that even possible or desirable? Yet perhaps the power in this statement lies in being declarative. I might not like it, but if such slogans drive action to save those species in need of our attention, then job done. Similarly, if by saving animals we end up saving plants, then who am I to complain about our furry friends attracting an unfair proportion of our attention. 'Charismatic-animal envy' is a debilitating infliction you may have noticed in many botanists, me included.

All this I tackle under the banner of 'observing and cataloguing nature' because that is what attracted me to botany and, as a sceptical botanist, I must be willing to kill – or at least interrogate – my own darlings.

Blood may flow over the wattle

When our beloved golden wattle almost became a Racosperma

A rose may still be *Rosa*, but gum trees are not necessarily *Eucalyptus*, and our wattles could soon be *Racosperma* rather than *Acacia*. Plant taxonomists (those who name plants) are breaking free from the shackles of tradition, and many well-known genera are being subdivided, resulting in a whole new suite of names.

It takes a great deal of patience and courage to learn the Latin name for a plant in the first place, so frequent name changes hardly

facilitate understanding and communication in the botanical or zoological world. Yet taxonomists have good reasons for the recent splitting of *Eucalyptus* into two genera,[2] and now *Acacia* into three, as well as the resurrection of an obscure name like *Racosperma*.

Sometimes changes are required due to the discovery of additional species that force a rearrangement of the existing classification. Other times, a reappraisal of the group, perhaps using new techniques such as electron microscopy or DNA sequencing, or new philosophies such as cladistics – a formalised method of arranging plants based on evolutionary theory – requires a radical rethinking.

Less constructively, changes occur when an older name is rediscovered, often after years of sieving through obscure and forgotten texts. The selection of a name for any newly described species or genus is governed by a set of strict laws, whose central dogma is 'priority'. In essence, the first name is the only correct name.

A 1986 proposal by botanist Les Pedley, from the Queensland Herbarium, to divide *Acacia* into three or more genera, means that names must be found for the new genera. Whether the genus is split or not depends on a value judgement based on all the available evidence. The correct names for the new, smaller genera, however, are determined by a code of law called the International Code of Botanical Nomenclature (ICBN).[3]

Although adherence is voluntarily, the ICBN is one of the few truly international codes of law. It consists of a set of rules for describing and naming plants which should be followed at all times and in all places (although a jury can be called in for tricky interpretations of the law). Botanists working in Australia, America, Europe and Asia all use the same name for a plant.

Pedley decided that dividing the wattles into three genera – *Acacia*, *Seneglia* and *Racosperma* – is the most useful way of classifying these plants. It also means that 96 per cent, or 850 species, of the Australian wattles will be included in one genus but, as things stand, this genus will not be called *Acacia*.

Following the ICBN, the correct name for this genus is *Racosperma*, the earliest available name for this distinctive group of plants which differs from acacias chiefly in having leaves that are actually modified stems called phyllodes.

But what if we want to call the Australian wattles *Acacia*? We happen to like this name, and it is widely used in forestry, horticulture and by all naturalists. What can we do?

The rules *can* be 'broken', but only in exceptional circumstances and following the equivalent of a court hearing. In this case, *Acacia* will still be used elsewhere (in Asia, Africa and tropical America) so the name is not lost entirely, just from our little corner of the world.

To keep such a name for the Australian cohort, or to devise a new but meaningful name (such as *Wattlia* or *Aussia* perhaps), would be contrary to the intent and principles of the IBCN. There are mechanisms allowing subtle changes to made in the rules, but removal of the idea of priority would involve an entire change in approach.

Rumblings along these lines were heard in Canberra recently where a strong contingent of taxonomists discussed a suite of radical changes to the 'Codes of Nomenclature' for both plants and animals. Ninety local and overseas biologists had gathered for a 3-day biotaxonomy workshop, titled 'Whose Name? What Specimen?'

The essence of the proposed change was a dilution of the priority rule. An 'approved list' of currently used names would be created, with those names not on this list discarded. Recent books and articles would be used as reference points for all accepted names, and a 'central authority' would pass judgement on all new names proposed for newly discovered plants or changes stemming from taxonomic revisions.

These officially sanctioned names would then be used throughout science and industry, and lead to greater stability for all concerned. It would save taxonomists considerable time and effort previously spent tracking down obscure names. Taxonomy would become more cost and time efficient, and those served by taxonomists would be happier.

Grand hopes indeed. Before we all rush out to embrace the new 'code', we must look a little closer at this enlightened vision. The heart of these reformers is in the right place (and their ideals would be supported by all working taxonomists) but can their methods work?

Who approves the names that appear on the list? Won't there be a huge problem matching up old and new literature on ecology, distribution, physiology and the like (and so keeping taxonomists wading through ancient texts)?

How will the 2 million or so described species be put onto an 'approved list'? Who has the time or resources to keep adding to this list (new species are described at a rate of about 30,000 a year at present)?

One participant at the workshop, Robert Andersen from DePaul University in Chicago, found at least two specific problems with the approach suggested.

It had been proposed that a future volume of the *Zoological Catalogue of Australia* series be used as a starting date for all the animals described in it. Andersen asked those present whether this text would provide the reference point for nonendemic species such as fox, dog, cat and rabbit. Most of those present felt that an international publication would be more suitable. However, the absence of a suitable world revision would seem to favour countries with recently completed Floras and Faunas (guides to the plants and animals of a region).

And for that matter, what if Australia 'snuck' *Acacia* into its Flora? Would the rest of the world have to wear it? How could regional bias ever be removed?

A second major worry arose from the issue of 'bad' taxonomy. An example was given in the Australian genus of blue-flowered herbs, *Brunonia*, where new plant names had been proposed based solely on the number of chromosomes in each cell. Most taxonomists disagree with this scheme, so Andersen asked whether these names would be included in an approved list?

The answer was a firm 'no'. Would the author of this proposal be allowed to add their names at some later date? Again, no. Whether one agrees or disagrees with the 'names by chromosome numbers', this is obviously a form of censorship. With no absolute means of knowing whether an author is wrong or right, such a system would seem unethical and counter to the spirit of the scientific method.

But back to the *Acacia* problem. Do we wince at the proposed changes for social or for scientific reasons? All name changes are a bother, but some are more bothersome than others. When little-known genera such as *Bassia*, *Baeckea*, *Cyperus* or *Monosporus* were split, there was little public outcry. However, when taxonomists start tinkering with *Eucalyptus*, *Acacia* and *Casuarina*, naturalists and taxonomists alike are aghast.

Arguments about the scientific and social merit of splitting genera such as *Eucalyptus* and *Acacia* have been thrashed about in the *Australian Systematic Botany Society Newsletter* over the last few years. Should factors such as commercial value, ecological importance and household names influence taxonomists? Taxonomy serves two major purposes: to provide practical floras and keys and to construct a useful classification system.

The taxonomists proposing the splitting of genera feel that both causes will ultimately benefit from their work, although the creation of a logical classification system must take precedence. The opposing camp feel that genera such as *Eucalyptus* and *Acacia* may be very large but they represent an easily recognisable group of plants. If they must be split, they say, do so below the level of genus – for example, into subgenera (noting that this is not always possible, with the science often requiring either the combination or splitting well-known genera).

One thing is certain: if *Acacia* and *Eucalyptus* are split into smaller genera, there will be lots of unhappy foresters, naturalists and gardeners. Whether we are all better served by these changes may take a generation or two of living with the new genera and their names.

The Age, May 1989[4]

Postscript

This is the oldest piece in this book, written when I was – for a year or so – a freelance writer for the science pages of *The Age*. It is a little more technical than most of the essays, and for that and other reasons unlikely to be published by an Australian newspaper today. I've included it here, with only a little editing, because it represents an interesting waypoint in resolving the nomenclature of wattles in Australia and South Africa (and elsewhere, but these two countries are where the controversy was most vigorous).

The eventual outcome was a rejection of an approved or sanctioned name list, and the splitting of *Acacia* into more than one genus – eventually five. As a result, the genus *Acacia* became more narrowly circumscribed, to include only thorny trees with fern-like leaves from Africa, America and Asia, with only a few species from Australia. The rest of the Australian species – all 950 of them, and the vast majority of what the world considered previously to be *Acacia* – were initially moved to *Racosperma*.

This decision led to fierce debate in the scientific community about whether to follow the rules of priority and accept this situation or to seek retention (conservation) of the name *Acacia* for the larger genus. Here are just some of the arguments made for and against both approaches:[5]

- If nothing further is done, 950 of the 1,300 species previously included in *Acacia* acquire a new genus name. Keeping *Acacia* for Australian species will minimise economic and social disruption.
- Despite the species numbers, acacias are ecologically, economically and scientifically more important in Africa. As it turns out, the number of books and papers that refer to *Acacia* in Australia compared to those referring to *Acacia* in Africa and elsewhere is about the same. This is perhaps surprising given the species number disparity but not really helping either side.
- Even if the name *Acacia* remains for use in Africa, around half of the African 'acacia' species will have to be renamed due to them being in one of the other new genera – that is, considerable change in Africa anyway while also reducing the amount of change required if *Acacia* is conserved for Australian species.
- Contrary to perceptions in Australia, the concept of an 'acacia' in much of the world is a thorny tree in Africa. Keeping the name *Acacia* for as many of these as possible would be widely supported by the general public, at least outside Australia.
- Exceptions to a strict priority of names is destabilising to the internationally accepted system of plant nomenclature. On the other hand, such decisions are taken from time to time in similar circumstances without an end to botanical nomenclature as we know it.

Considering this, and more, a decision was made at the 17th International Botanical Congress held in Vienna in 2005, to 'conserve' the name *Acacia* for all but a handful of Australia's nearly 1,000 wattle species, along with a few from Asia, some Pacific Islands and Madagascar.

The South African 'type' (the species and linked preserved herbarium specimen chosen to represent the genus) was replaced with one from Australia, and *Acacia* became a *nomen conservandum*, meaning the name has been formally approved for conservation against earlier names by the General Committee of the Nomenclature Section at an International Botanical Congress.

Unfortunately – for nomenclatural stability and perceptions – that decision in Vienna was considered approved even though only supported by a minority of those voting. This is because it came recommended by an expert panel, which under the agreed rules meant a 60 per cent majority was needed for its rejection. Six years later, at the next opportunity to test the outcome, a more respectable 70 per cent of participants voted to support the 2005 decision. Unfortunately – again for perceptions of fairness – that 2011 meeting was held in Australia, which had the most to gain by the change.

The decision continues to be challenged and remains contentious today, as Natasha Mitchell and Belinda Smith found when they covered the topic for ABC RN's *Science Friction* in 2021.[6] A pertinent observation by Australian botanist Dr Kevin Thiele at the end of that report is that these decisions only refer to the use of scientific names in the scientific literature. While it is important in science to have consistency and universality, outside this system any name can be, and does get, used.

Thus, in Africa, acacia rather than *Acacia* will continue to be used for their emblematic thorny tree, while in Australia, most people will always call our ubiquitous, yellow pom-pom-flowered bush a wattle.

You say boab, I say that's OK

For common names, anything goes

Amid the excitement of the public poll for Australia's favourite tree in 2022,[7] it became clear that the common name 'ghost gum' was being used for a variety of eucalypt species with white, ghostly bark. This

meant the ranking of *Corymbia aparrerinja* as Australia's third favourite tree will forever be doubtful.

No big deal, but this is an almost inevitable consequence of using common names. There are no rules. Despite what some might say, there is no right or wrong, no cause for a pedant to correct your pronunciation and no need ever to use a different name. That is part of their charm.

While a scientific or botanical name must obey some rather austere rules of nomenclature, a common or vernacular name can be whatever you like. Sometimes, that leads to plants in different species being called the same thing, as happened in that national poll.

There are other good examples. In Tasmania, the blue gum is *Eucalyptus globulus*, in Sydney it is *Eucalyptus saligna*. Of course, when communicating across State borders, you can always add the helpful 'Tasmanian' or 'Sydney' to that common name.

You might even find that you end up using an all-purpose name for a few different things in your neighbourhood. My father seemed to call any evergreen shrub with glossy, green leaves, japonica. Others would restrict this name to the flowering quince (*Chaenomeles japonica*) or perhaps the common camellia (*Camellia japonica*). Usually, that doesn't matter.

Unless, perhaps, you intend to consume the plant for food or medicine, or perhaps take action on what seems to be a threatened species or a potentially weedy plant. You will also find common names a hinderance when communicating with someone who speaks a language other than English. So yes, there are good reasons for botanical nomenclature.

Often though, a common name is fine. While Floras and other expert guides to the plants of your region may recommend a common name or two, there is no need to take any notice of them. At their best, common names are generated locally and evolve to suit those who use them.

Take the London plane, that resilient street tree that irritates (in all senses) many. This common name is used in Australia for hybrids between the oriental plane (*Plantanus orientalis*) and the American sycamore (*Plantanus occidentalis*) and is these days given the botanical name of *Plantanus × hispancia*.

Some (well-meaning) pedants apply the name London plane more narrowly, to a clone of *Plantanus × hispanica* introduced into Britain in the 17th or 18th centuries, perhaps via the Oxford Botanic Garden.[8] I am told all the plane trees in Melbourne are a different clone and unlike the older trees revered in the streets of London. Either way, all are hybrids between a species from the eastern Mediterranean and one from eastern North America, most likely crossed somewhere in southern Europe. London plane indeed.

The etymology of the baobab tree is fraught too. This common name was devised and used in Madagascar for species in the genus *Adansonia*, including what many in Australia would call a bottle tree, or, more controversially, a boab. Note that the unrelated *Brachychiton rupestris*, from inland Queensland, is more commonly called bottle tree and mistakenly (although call it what you like) boab.

The baobab or boab genus, *Adansonia*, was created – to be clear, for scientists by scientists in their mutually agreed scientific language – in 1762. Its author was the Swedish botanist, pharmacist and, 9 years earlier, the creator of the system of biological nomenclature in which this name was established, Carl Linnaeus.

The name *Adansonia* honours French botanist, Michael Adanson, who, in addition to publishing the first full botanical description of the boab, spent much of his life unsuccessfully urging scientists to adopt his rather than Linnaeus' system of classification for plants (inspiring to some extent a recent novel by David Diop, *Beyond the Door of No Return*).[9] Linnaeus was happy, it seems, to honour his rival but not inclined to use the name suggested by Adanson, '*baobab*'.

The baobabs had been marvelled at, and written about, by Arab traders and European explorers in the 16th century, and their fruit described by Venetian botanist Prosperous Alpinus under the name bahobab – probably drawn from the Arabic word *buhibab* for 'many seeded fruit'. That many-seeded fruit made its way to Egypt, and so to European awareness, although it was only when Adanson visited Senegal and provided his botanical description and illustration, that Linnaeus was prompted to add this group of plants to his botanical schema.

It was botanist and fleetingly director of Sydney's botanic garden, Alan Cunningham, who in 1827 published the first written account of the *jungeri* (among other local names for the boab) in Australia. It took another 30 years for the connection between these trees and those of Africa and Madagascar to be made, by the inaugural director of Melbourne's Botanic Gardens, Ferdinand von Mueller. Mueller described the Australian species as *Adansonia gregorii* after encountering the species on the Gregory Expedition to northern Australia.[10]

Our single Australian species of *Adansonia* is today found 'naturally' only in the Kimberley region of northern Australia. The genetics of Australian populations point to its origin being a seed of the African species, *Adansonia digitata*, that drifted across the Atlantic from Madagascar sometime over the last 6 million years. This was after Australia separated from Africa and Madagascar in the split of Gondwana, but most likely before the First Australians arrived, some 60,000 or so years ago.

Once established in the Kimberley, *Adansonia digitata* seems to have adapted to the new environment, and over a few million years evolved into a distinct species. From its genetics, coupled with linguistic studies, it is most likely that the species spread further inland thanks to First Australians carrying seed with them for food and celebration.[11]

Most English-speaking people in the Kimberly now call *Adansonia gregorii* boab, but there are of course many more Indigenous language names[12] such as *jungeri* (Wunambal language group), *potkurri* (Wunambal), *larrkari* (Bunuba) and *katawun* (Miriwoong). Linguists have found a strong correlation between the evolution of these names, and changes in the genetics across the range of the boab, concluding that the spread of the species eastward was thanks to the First Australians.

Given we now have an Australian endemic species, the name boab seems appropriate. Yet some argue that because Madagascar is where most species of *Adansonia* reside and because the name baobab originated from a local language there, everyone in the world should use this common name rather than corruptions such as boab.

I disagree with this logic. As I've said, a common name can be whatever you like and the best of them reflect community usage. In this case, Australians have – as they do – shortened the name to something a little pithier. We are just lucky it's not boabo or boabie (as in 'arvo' and 'ciggie'…).

So, if you prefer baobab, go right ahead. Don't fret though when Aussies use boab or even bottle tree.

Gardening Australia, March 2024[13]

New species are where you find them

Sometimes it is easier to not describe a new species

I enjoyed reading Caroline Cook's *Tell Them to be Quiet and Wait*,[14] a warm-hearted and well written novel constructed around the life of American phycologist, Hanna Croasdale, born in 1905. A phycologist is someone who studies algae, like I do, and Croasdale is a dominant name in the literature I use to identify algae in Australian streams and lakes. I never met her but have colleagues who did.

The fictional Beverley Conner, an aspiring academic based on Dr Croasdale, struggles for recognition of her talent, for a lasting career in phycology, and for equitable recompense in a male-dominated American college. Sadly, it's a story with parallels still today, and Cook shows this by interspersing Croasdale's story with that of a woman entering college in the current era.

One detail was less convincing though: the author's obvious reverence for the discovery of a new species. While this is an exciting and fun thing to do, I would not consider the finding of a new species the measure of a good taxonomist (someone who names and describes

species). In Cook's novel, Beverley Conner dismisses gentle podding from her partner Rachel about not getting due recognition from her colleagues when she discovers a new species (of algae) by saying it is 'what all scientists do'. Rachel responds: 'It's not. It's what good ones do.'

Rachel is the voice of reason in this story, the implication being that finding a new species is a test or even proof of being a good scientist. This is a contentious matter in the novel because few others in the department have discovered new species yet they have a higher standing than Conner.

While its clear Conner is woefully under-recognised and recompensed, I take issue that the naming of a new species has much significance. While I'm a bloke, and things were still stacked in my favour in the 1980s, I was awarded a post-doctoral fellowship on the basis of one published taxonomic paper and no new species. A new species wouldn't have changed that outcome, nor should it have.

By the time I was employed in an ongoing role as a taxonomist, 4 years later, I *had* published four new species and a new genus, yet I'm convinced it was the rigour of my science not the tally of species that impressed the selection committee. I was told as a post-graduate, and believe it true still today, that it is far easier to describe a new species than, after much study, decide none is required. This doesn't sound as exciting, right? But that's the way it is.

As my career progressed, I eventually did describe many new species. And genera. And one family. Fifty overall, along with about the same number of what we call transfers, when a name is moved from one category to another – a reflection of newly discovered relationships between groups of species.

But I retain my scepticism of the discovery of new species as any kind of taxonomic intelligence indicator. First, there's the finding of the thing. That might be done by the phycologist themself, or by someone else. Either way, the organism has most probably always been where you find it – at least for a few thousand years of evolution. Or if you find it among existing collections, then again it was there 'to be discovered'.

There is much luck involved, albeit with a tendency to explore places previously not explored by others of your kind (phycologists, for example) helping. Admittedly a good eye is important, so training or expertise will mean that you should find something others won't. That can be in the field or in the laboratory – sometimes more often the latter, if you are looking for tiny algae in water samples, as I expect Dr Croasdale did on most occasions.

When you finally spot a species that is new to you, and potentially new to this location, the fun begins. You either shoe-horn it into an existing category or decide it is new to science and prepare to name and describe it in the scientific literature. The latter is evidently what Connor/Croasdale did.

In my experience it is definitely easier to describe something as new. Just whip up a Latin or Greek name that explains or entertains, describe its key features, and publish. Ideally your description is what we call a 'diagnosis', comparing the new species to existing ones, but (poor, in my opinion) taxonomists don't even have to do this.

Sometimes, a new name is more of a flag in the ground to indicate there is something interesting here to be sorted later. Often, more often than you might think, a name is later considered unnecessary, and the 'taxon' subsumed within another. This is usually when new evidence is found, or the group becomes better understood by taxonomists. In any case, if you publish in a non-refereed journal or make your case boldly enough in a peer-reviewed one, it's not very hard to publish anything as a new species.

If you decide *not* to describe your interesting find as a new species, you are saying the 'discovery' is the same as something recorded from elsewhere – based on a thorough review of published accounts, preserved collections and sometimes further collecting in areas to improve understanding of the variation in your species or existing ones. Your discovery can often change the circumscription or concept of the species in question, all part of the hypothesis testing inherent in taxonomy. This is a complex process and there is no shame in it. It can show more courage and fortitude *not* to consider your species as novel.

Then there is the ease of finding new species in some places and for some groups of organisms. You are more likely to find a new alga than a new flowering plant just because taxonomists have not yet look as widely for them and there are fewer amateurs combing the Earth and finding species with limited distributions.

This is not to say we don't get just as excited by news species in the world of phycology. I often tell the story of *Entwisleia bella*, that genus of red seaweed named after me by colleagues (and friends) Fiona Scott, Gary Saunders and Gerry Kraft. (I always add that the species name *bella* is a reference to the beauty of the alga, not me, and a nod to I think Fiona's Italian grandfather, who ignited her interest in the sea and seaweeds.)

Not only is this alga a new species and genus to science but its discovery demanded the creation of a new family (Entwisleiaceae) and order (Entwisleiales). This means Fiona finding *Entwisleia* on a subtidal rock platform just south of Hobart, in Tasmania, was like the first scientific sighting of a conifer, primate, spider or diprotodontids (kangaroo, possum, koala or possum) on Earth. Those groups of organisms are also orders.

Anyway, the discovery goes to show how much there is still to discover in Australia's algal flora. I got into algae myself after finding something unusual and new for Victoria when collecting in Melbourne's Darebin Creek for a 3-year university project. It was a red alga, like *Entwisleia*, and I spent most of my research life working on an order of red algae called the Batrachospermales. These red algae are *not* found in the sea.

Yet it was the superficial resemblance between this subtidal alga and the Batrachospermales that led to it being named after me. While it looked similar, DNA sequencing showed it to be quite unrelated to these freshwater red algae and even to its companions in the ocean.

From all this you can see there is plenty of serendipity in first finding new things (to science) and then deciding how important that finding is. For Dr Croasdale to find a new species would have been a real buzz, but not necessarily proof of wisdom.

I should end with a warning about using the word 'discovery'. We used to say that James Cook and his crew 'discovered' Australia. Clearly the land had always been here and there were others living there before Europeans arrived.

For a taxonomist, the best you can do is be the one who finds a species 'new to science', given it similarly has presumably been there for some time and may well be known to other humans without its being represented by a Latin name and pithy diagnosis.

April 2024

Enjoy nature kids, but keep your gadget handy

It would be short-sighted not to use technology as a way to complement and interpret the natural world

Having just read another 'ditch the device, live your life' story in a weekend paper I'm pretty much over it. Not the literal device but the literary device – worthy opinion pieces on the evils of the virtual world and the riches to be reaped in the real.

What a time to dissent. Australia's first Nature Play Week, launched at Royal Botanic Gardens Melbourne on 1 April 2014, is all about encouraging kids and their parents to take a break from the screen. The Kids in Nature Network want our young ones to spend more time outdoors, experiencing and enjoying nature.

I agree entirely, and that is the dilemma. Nature is good, but so is eNature (or iNature, if you are that way inclined). I want kids to be able to enjoy the real world made better by the virtual world. I'm not convinced by opinions such as those attributed to English Professor of

Synaptic Pharmacology, Susan Greenfield, that gadgets and technology are changing the way we think, and not in a good way.

Unlike Professor Greenfield I'm no expert on brains, but her arguments and evidence to date don't convince me as a parent or a scientist. You can find learned refutations of her views on the web.[15]

My own childhood 'device' was the book, although through necessity rather than choice given the internet wasn't invented and television was rationed by my parents. Book reading is still a habit, on paper and on a device, but I don't see it as some virtuous pursuit. A good novel, dare I say, takes us to a virtual world, and if good enough, will distract us from those pesky personal interactions that we apparently need so desperately.

I am not writing in defence of obsessive online gaming or social networking. I don't do the former, just like I don't play cards or board games. As for the latter, the less social the better for me (Twitter[16] is perfect) and I'm happy dropping into these platforms like a pub or a dinner party, leaving when I've had enough.

The brilliant author and illustrator Graeme Base, who is launching Nature Play Week for us, will have kept many a child (and adult) from playing in nature. Of course, his works will have also coaxed many a reluctant child out of doors in search of strange beasts or to create their own worlds from sticks and mud. Or perhaps from pixels on a screen.

A good book, like a good website or social media post, may at first detract or distract, but at its best it adds to the world around us. Tweeting, Pinteresting or Instagramming an image from the botanic gardens is a delightful way to share your joy, to encourage others to visit and remind them that plants are both pretty and interesting.

The great museums, galleries and gardens of the world all provide terabytes of online information accessible through mobile phones or your home mainframe. Royal Botanic Gardens Melbourne has an app in development to help you find and identify plants in our gardens, and then find out more about their classification, biology and propagation.[17]

This week, as every week, I will be encouraging people of all ages to walk on our grass, hug our trees and smell the flowers in Melbourne and Cranbourne's Royal Botanic Gardens. I will also encourage them to search our website for stories and facts about plants, and to join us on Facebook and Twitter as we share the fun of their visit.

Because it's Nature Play Week, I'll make a special effort to put my gadget on silent, slip it in my bag or perhaps even leave it at home for a day. But when I want to find out more about a bizarre bloom, the latest botanical discovery or the likely impact of climate change on our gardens, I might just take a peak if I may.

Enjoy nature, live your life, but stay connected.

Unpublished, March 2014

Postscript

This essay was never published but later the same year, I expressed similar sentiments to *The Age* journalist Gina McColl.[18] McColl was writing an article on criticism she had received about the Disney-themed 'fairies trail' running at Royal Botanic Gardens Melbourne. This screen-based trail featured Disney fairies and was, said critics, 'compromising the garden's integrity and incorporating stealth marketing to children'.

The software included in-app promotions, which one academic condemned as deceptive marketing given the profile of the Gardens as something 'non-commercial'. I could bear but debate that accusation, a perennial tension in a place partly depending on commercial income.

Evidence about screen time damaging children's health and development was also cited, including its effect of turning play into an individual rather social experience. I'm sure this is true, to an extent. I get the feeling that small results are extrapolated well beyond their application, but again, I'm no expert in this field.

What I *did* see was happy children enjoying the Gardens with their family and sharing their experiences with equally happy parents. I saw the app as a gateway to nature, not a gate blocking access. I was quoted as saying 'Children [are] using screens and accessing the world through them, that's the reality. You then try and use that to get them to connect with nature. If you can find clever ways to do that, all the better.'

This was aside from the educational messages our staff added to the app, with fairies guiding children to a deeper understanding of what a tree is and what it does. I remembered one mother telling me that her daughter was allowed an hour or two of screen time a day, and she was delighted that her screen time allocation today was mostly outside in the botanic garden.

The argument I had no truck with was the anti-Disney one. The Walt Disney Foundation was a partner in this education programme and had donated the software and iPads we made available to visitors on arrival. It wouldn't have happened without their support so yes, we were openly promoting both a supporter of the Gardens and the producer of popular childhood characters. Both of which we had done many times before.

In response to criticism of our support for a company like Disney, I said, 'We're very much a May Gibbs and Norman Lindsay kind of place rather than Disney and Pixar, so [in this instance, we are] really trying not to be too elitist about who comes to the gardens and encourage an audience who might not normally know that much about us.'

The Disney fairies were much loved and legal, and I didn't feel it was my place to put value judgements on good taste. One commentator was not fussed about using apps in the Gardens but distraught – using words such as 'outrageous' and 'travesty' – about the Disney fairies. It showed a lack of creativity, he said.

Well, perhaps. But creativity has a cost, and again I would counter that kids actually like the Disney fairies. Any concoction we might devise with our own budget would be worthy but unlikely to compete. The critic complained that what we had rolled out could be run anywhere was all about 'international commercialisation'.

This is such a big topic I don't quite know where to start. Let's just say that the Gardens will always do a mix of things that represent its place in Melbourne, and Australia, and other things that are part of it being in a community with an international purview. I can't help but think that putting on Shakespeare, showing international films at Moonlight Cinema and promoting the occasional flowering of the titan arum (*Amorphophallus titanum*) from Sumatra are all comparable exercises.

In any case, I think it is elitist to ban Disney fairies from the Gardens, short-sighted not to use technology to complement and interpret the natural world, and unless the community and its governments wish to subsidise the enterprise in full (and in lieu of an entry fee), entirely reasonable for the Gardens to have commercial arrangements with benefits accruing to the sponsor or commercial partner.

I accept extinction, reluctantly

When we chose to save one species, another goes extinct

'I will not accept extinction', proclaims Carlos Magdalena, repeatedly, in his 2018 memoir, *The Plant Messiah*.[19] Part posturing I'm sure, but apparently a strongly held conviction by my former colleague at Royal Botanic Gardens Kew.

Yet species *do* go extinct. In 3.8 billion years of evolution, planet Earth has lost uncountable numbers of them, and will continue to do so. Evolution depends on a species being transient, including *Homo sapiens*.

What most people, including my friend Carlos, mean is that they want an end to 'human-induced extinction', or perhaps more precisely, the extinction of charismatic flora and fauna known to be alive on Earth today.

Extinctions are certainly more frequent today than at almost any other time in history. It has been estimated by some that over the last century, species have been lost at a rate of 100 to 1,000 times the historical average, thanks largely to pernicious and persistent interventions by humans.

In addition, we shouldn't discount the special significance to us of the species alive today. These are the plants, animals and other life we evolved with as a species, the ones we depend upon and share our planet with. They are also the ones whose existence we *can* influence – for better or worse.

Perhaps no known species should go extinct on our watch. This could be argued from a utilitarian perspective. We don't know which species are necessary for our survival and it is a courageous step to lose any of them, let alone try to predict which ones might lead to severe consequences for our own well-being.

If we accept that some extinction is inevitable, though, which species might we protect as exceptions? Those that benefit humans? Those that keep our forests healthy? Or those that are the most beautiful or odd?

Some people argue for maintaining as much genetic and evolutionary diversity as possible, keeping all the building blocks we might need in a crisis and avoiding 'benefits forgone', to use an economic term: all those potential life-saving drugs, new foods or cross breeding opportunities to adapt to climate change, for example.

As for the rest, what might appear to be redundancy may turn out to be exactly what is required for an ecosystem to avoid what are called 'tipping points' – when things change dramatically rather than gradually, due to some externality. We know that many species are not able to adapt or move outside the ecosystem in which they evolved, so this 'adaptive capacity' or resilience of the system as a whole may be critical.

On top of these practical considerations, each is a unique outcome of evolution and will not evolve in the same way again. As Steve Hopper from Royal Botanic Gardens Kew once said, the extinction of a species is the equivalent of losing Da Vinci's *Mona Lisa* forever.

If that Mona Lisa now only lives on in a zoo or garden display, it has an eery parallel with a painting in an art gallery. These 'living dead' – species that are critically endangered and survival is dependent on humans intervening in some way – are either something to give us hope or to feel depressed about, depending on your world view.

Outside the lecture theatre where I was delivering a speech on this very topic in 2009, grew a cultivated specimen of *Eucalyptus copulans*, a reminder of a species clinging to existence. At that time there were still two individuals remaining in 'the wild', but its future very much depends on what we do next.

So too, the Wollemi pine, clearly on its way to extinction when it was brought to the attention of the scientific world in 1994 by a National

Parks ranger. The fossil record shows it once covered vast areas of Gondwana but now there are fewer than 100 trees surviving in a small collective of canyons not far from Sydney. A single bushfire[20] or pest incursion could wipe them all out.

It could be argued that we should give the Wollemi pine a nudge over the (extinction) ledge and put any resources saved into other species behind it in the queue. It doesn't seem to have been humans that led to its initial rareness (although its possible fire frequency over many thousands of years played a part) so saving this species might perhaps, be considered less of a moral imperative. Different to, say, Victoria's Leadbeater's possum, the orangutang or the panda, where we are clearly responsible for their demise.

But it doesn't feel right, does it? To abandon a species tinkering on the edge of extinction while we have (potentially) the ability to stop its demise. Should we ever knowingly accept that a species is about to go extinct? Maybe not, but every time we invest resources (time, energy, money) to save one species, others will suffer.

This dilemma is captured in the concept of conservation triaging, where the limited resources available for conservation are allocated to *maximize* returns. Decisions are made by explicitly accounting for the costs, benefits and likelihood of success of alternative actions.

Not everyone agrees with this approach and conservation triage is viewed by many as a disincentive to increase conservation funding and innovation. It is labelled as 'defeatist' and an unacceptable trend towards accepting extinction rather than the conservation of *all* biodiversity.

There are also doubts about the comparison made between conservation and emergency medicine, where the concept has currency. Resources for a medical emergency are fixed, it is argued, while for conservation they are not. Also, while in medicine there is universal agreement on the moral value of patients, the objectives in conservation not universally agreed.

Interestingly, the COVID pandemic tested whether there really was universal agreement about the value of life, with the elderly dying in greater numbers and some jurisdictions deliberately favouring the young when healthcare resources became constrained.

I can't help but conclude that we do triage resources, whether or not we admit it. The simple creation of Red Lists and categories of rarity and threat does this. It is hard to avoid the government and community focusing on those most at risk rather than those most likely to be saved.

Still, as one opponent concluded, 'conservation triage's greatest contribution may be as a vehicle for demonstrating that scant resources is not the greatest threat to conservation. Rather, the greatest threats to conservation are values and policies that are antithetical to conservation.'[21]

April 2024[22]

Eager for spring in London

Is it an early spring? Yes and no

I was out kayaking on the Thames in Richmond on Monday. I had feeling in about half my fingers, so I figured it was a particularly mild new year. That's fine for bragging when I return home, but it's hardly climate science.

Strange things happen when we get unusual weather. We start to observe things, sometimes for the first time, and we see patterns when none exist.

The sum of our collective observations right now seems to be that we are having an early spring. It wasn't good enough that bluebells came a month early in March last year; now we have a whole season moved 2 or 3 months forward. But as George Monbiot has confirmed,[23] while our autumn was the second warmest on record, December was only a little warmer than average.

Actually, I have no problem with extreme extrapolation, as I like to call it. Anything that messes with the seasonal system is grist to my mill. The four-season system doesn't work at all for much of the world yet it continues to be followed dutifully in almost every country.

We need more observations and debate, not less. I work at Royal Botanic Gardens, Kew now, but until 9 months ago, as director of Sydney's Botanic Gardens, I enjoyed a far less seasonal climate in Australia. Although more subtle, I was convinced Sydney experienced at least five seasons every year.

It seemed that after a long, 4-month summer there was (only just) a 2-month autumn, followed by an equally short and mild winter. Then the action started with an early spring in August and September.

Here in London, I'm noticing far more drama in the seasons. I don't have enough experience yet to throw my support behind the 'Vivaldi seasons', as I call the classic four, but I am more than a little intrigued by a five-season model proposed in 1954 by UK meteorologist Hubert H Lamb, in his charmingly titled book, *The English Climate*.

In my first winter, I'm finding things a little messy. What are the flowers telling us? Is it an early spring? Well, yes and no. It seems we have a little early spring, a little late summer and a big dollop of normal winter. Some plants are flowering when they shouldn't while others are doing what they should.

A good example of the latter is the winter cherry, which flowers every year ... in winter. Snowdrops may be out early but they are always winter flowerers. It is true that we've only had a couple of frosts so far, so the flowers and fruits of some tender plants have lasted longer than you might expect. I have a fuchsia-flowered gooseberry crammed between a bench and our south-facing wall which is, unexpectedly, in full flower and leaf.

My colleague at Kew, Andy Jackson, is much more familiar with British winters. He reckons there are three groups of plants in flower at the moment. The first are the winter-flowering plants. These are the ones that shouldn't surprise us but perhaps do because we are not used to observing properly. In this group are heather, hellebores, wintersweet, *Viburnum* and camellias.

Then there are the 'early flowerers', some of them up to 4 to 8 weeks earlier than expected. These are the ones that beg for the 'early spring' headline. In here are daffodils, primrose (although often an unseasonal flowerer), *Rhododendron lutescens* and *Colletia spinosa*. You can add my fuchsia-flowered gooseberry.

Finally, Andy lists a bunch that have continued flowering from the late summer through into winter: plants such as *Cyclamen hederifolium*, *Lavatera*, *Abutilon* and rosemary.

You get the idea. There are always plants in flower, even in the middle of winter, but it seems we have more this year than usual. Some are early and some are late.

On the weekend, I tested this out at Savill Garden, near Windsor. I found one early daffodil, but mostly the botanical displays were what you would expect to see in winter: wintersweet flowers, dogwood stems, camellia blooms, winter cherry blossom, the odd rhododendron (azalea) in flower and lots of hellebores.

As for that daffodil, does it a spring make?

The Guardian, January 2012[24]

Postscript

I wrote this essay in mid-winter while editing the final pages of *Sprinter and Sprummer: Australia's Changing Seasons*.[25] Over the ensuing months, I wrote these two shorter pieces (both unpublished) on the perspective of Londoner's to the 'real' arrival of spring.

Nothing to see here, just spring again (*London, 1 March 2012*)

It's finally spring. Well, it is if you use the first of the month to kick-start your seasons or as many pundits do, because the 'tommies' are out. If you prefer your seasons to be based on solstices and equinoxes you have 20 or so days to wait.

As a botanist I'm happy to pronounce that spring has arrived. In the last few days there have been plenty of signs. About half the daffodils lining Kew Gardens' Broad Walk are in flower. The Rock Garden and other bulb collections are erupting daily with new blooms. And I saw my first tulip today, a lovely cream and purple bloom of *Tulipa kaufmanniana*.

As for *Crocus tomasinianus*, the tommy, it has been in flower since the second week in January. Granted that was indeed the first flower, but I reckon they peaked a few weeks ago. I blogged about them on 26 January, taking great pleasure in reporting that this so-called early spring flower blooms, on average, on that very day every year. This year, though, it was earlier than usual.

We know this because at Royal Botanic Gardens, Kew we track the first flowering of a hundred different kinds of plant in *Kew100*.[26] The early crocus, as this species also commonly called, flowers, well, early every year.

Thanks to the work of a team of volunteers lead by the Kew Wildlife and Phenology Officer, Sandra Bell, there is excellent data for the last 10 years or so, building on some solid observations back into the last century. ('Phenology', by the way, has nothing to do with assessing intelligence using bumps on your head but a lot to do with when plants flower, fruit and do other seasonal things.)

The days of the year are numbered 1 to 365, or in a leap year like this one, 1 to 366. I have access to a little more of the information behind the website so I could announce that day 26, 26 January, is the mean of all the measurements we have for the early crocus. That is, this is the day you would expect the early crocus to start flowering in most years.

This year we had a particularly early start, with the first flower reported on day 9, 9 January. We were only one day off the earliest on record since the 1950s, recorded in 2007. By 26 January the early crocus was well advanced across Kew and by early February in peak flower (by my estimation).

Despite these two recent early flowerings, there isn't really a trend in Kew's data yet, so we can't say if climate change is influencing the early crocus. But there is good evidence in studies elsewhere that spring flowerings are up to a few weeks earlier than they were a century or so ago.

As to celebrating the start of spring, my advice is not to wait for the Equinox. Visit your local garden today, tomorrow, and as many days as you can so that you don't miss the parade of spring flowers, whether they come early or late.

Bluebells chime with climate change (*London, 1 April 2012*)

This year at Kew Gardens the bluebells flowered in March for the first time since records began. Like the tommies, they were weeks earlier than the long-term average. The opening of the bell-like blooms on 29 March marked a new milestone for the insidious 'spring creep'.

It wasn't so long ago that Kew held a Woodland Wonders festival in May to celebrate the bluebells blooming. Now celebrating spring is a little like working out when Easter will be held, without the quirky lunar formula.

This year's first flowering of the bluebells is 6 days earlier than the previous record, held by the years 2002 and 2007. Back in the 1980s and 1990s, the average date for the first flowering in each decade was 25 April and 21 April. The march (excuse the pun) had begun.

The decadal average for the new millennium is 13 April and slipping ever closer to the start of the month. It is now 20 years or more since the bluebells started flowering in May, and no matter how bleak our winter or spring we don't expect to be holding a bluebell festival in that month ever again.

The string of warm sunny days since mid-March have brought forward lots of flowering shrubs as well – things like *Prunus* 'Kanzan', *Buxus sempervirens* and *Malus floribunda*. The magnolias have come and almost gone at Kew Gardens. The boastfully named glory of the snow (*Chionodoxa*) has been carpeting one of our lawns in blue for weeks and the fritillarias are about to hit their peak.

Clearly there is lots happening all at once and we expect a concertinaed as well as early spring. None of which proves climate change is occurring or caused by humans. That was demonstrated beyond doubt long ago by climate change scientists.

Still, changes to flowering and fruiting times have been well corroborated with seasonal observations over hundreds of years. Grape vines in Burgundy now ripen up to 4 weeks earlier than when Philip of Rouvres, Duke of Burgundy in 1354, downed a goblet of local wine.

Spring creep, as it is called, has already caused mismatches between organisms that rely on day length and those that respond to temperature. Caterpillars in the Netherlands are peaking 16 days earlier than in 1985, leading to 90 per cent decline in population of the migratory pied flycatcher that now arrives a few weeks too late to feast.

The hungry caterpillars presumably munch their way through more plants as a result. Elsewhere, flowers and their pollinators are likely to become increasingly out of sync, leading to failed seed set and harvests.

Our gardens, flowers, insects and animals are all telling us something. The bluebells are ringing in change, whether we like it or not.

Australia should scrap the four seasons

We need not only more than four seasons in Australia but more than one system of seasons

September is considered the start of spring by most Australians but I think we have it all wrong. In the south at least, we should be

celebrating an 'early spring' in August – when the wattles are blooming *en masse* – and a 'late spring' beginning in October.

As I argued in my 2014 book, *Sprinter and Sprummer: Australia's Changing Seasons*,[27] the seasons here are misunderstood, misinterpreted and misused. Why should we have four of them? Why must they each take up 3 months of the year? And why in Australia are the Aboriginal seasons so variable in number and length?

Living in London for 2 years, I gained first-hand experience of 'true seasons' and of the plants that either define them or respond to them, depending on your perspective. I found that even in England, the four seasons don't always match the annual cycles of nature.[28]

Yet seasons have been with us since early in recorded human history. In fact, we could argue that many animals track seasons for breeding and feeding, so our species, *Homo sapiens*, has merely extended this concept a little. Typically, by dividing the solar year into segments that start at predefined times.

In ancient Mesopotamia, the year was divided into two: one half beginning with the sowing of the barley (autumn), the other with its harvesting (spring). Early Egyptians, living beside the Nile, added an extra season and brought explicit reference to cold and hot seasons: 'flood', winter, and summer.

The 'Vivaldi Option' of four neatly defined seasons appears to have originated in the Mediterranean region, though it may have also emerged independently in China and in places with less well documented histories. The Sumerians and Babylonians were the first in the region to use equinoxes and solstices to define four evenly timed seasons.

This four-season system was taken up by the Greeks, and then by the Romans, and so it spread through Europe and eventually to colonial countries such as Australia. The four seasons start regularly on the first day of a month, or sometimes the 21st or thereabouts, depending on local habits and quirks.

Others before me have suggested this makes no sense in much of Australia. But as with the Union Jack in our flag or the monarchy, we seem reluctant to change. In the 1990s, environmental educator Alan Reid encouraged members of the Gould League of Victoria to record their seasonal observations as part of his 'Timelines' project, resulting in a six-season proposal for Melbourne.

This pales beside observations over tens of thousands of years by Aboriginal people, resulting in finely honed systems of two to 13 seasons (I found only one example of four seasons, with six is the most common number), depending on their particular Country.

Wary of appropriating Indigenous knowledge, in my book I proposed a tweaking of the European system. The familiar anchors – summer and winter – are still there but with more granularity in what lies between. I argued this system was a better match for what I was observing in our natural world, particularly the plant world.

The result was a five-season system for all of Australia south of Brisbane:

- **Sprinter** (August and September) – early Australian spring and the start of my seasonal year. This is when the bushland and our gardens burst into flower. And also when that quintessential Australian plant, the wattle, is in peak flowering across most of Australia.
- **Sprummer** (October and November) – the changeable season, bringing a second wave of flowering, particularly for trees.
- **Summer** (December to March) – extending into March, when fine warm days continue in southern Australia. A subdued season for plants.
- **Autumn** (April and May) – barely registering in Sydney, but needed further south where there can be good autumn colour on exotic trees. This is also peak fungal fruiting time.
- **Winter** (June and July) – a short burst of colder weather, when the plant world prepares for the sprinter ahead.

The first season, sprinter, is my most important change. It is easy to recognise and backed up by good observational data from nature and preserved herbarium specimens (as I demonstrate further in my book). The other four seasons are perhaps more aspirational: concepts to test and probe a little further.

Then there is climate change, and the fact that the seasons are changing, whether we like it or not. Perhaps we need an evolving system of seasons. However, I argued, we should at least get our seasonal house in order first.

There are no 'perfect' or 'correct' seasons. I am happy for my system to be rigorously debated and tested, and I would be thrilled if, through further observations and monitoring of the natural world by others, the system is entirely redesigned.

I'm aware, too, that I've relied on conjecture and perception alongside peer-reviewed evidence, of which there is precious little on the seasonality of Australian plants and animals. Thankfully that has changed, a little, over recent years, with more studies driven by the urgent need to address climate change.

When I was writing *Sprinter and Sprummer*, over a decade ago now, I was convinced that most people didn't care about which seasons we use or what we call them. No-one seemed to have responsibility for seasons – other than, vicariously, the Bureau of Meteorology – and in any case, I didn't expect Australia's seasons would ever be changed.

That still holds true today, but 'most people' doesn't include Indigenous Australians. As the country's primary knowledge holders, Aboriginal and Torres Strait Island peoples should lead the consideration of any new seasonal systems.

I am also aware that seasons are not the same across southern Australia, varying with latitude, latitude and distance from the sea, for starters. My system, then, is a very rough first cut to get people thinking. To start a conversation if you like.

Since the publication of *Sprinter and Sprummer*, I've given hundreds of talks and interviews, with mixed but mostly positive reception. Mostly, people like the idea, particularly the early spring (sprinter). But that's where it ends.

In time I hope Australians will seek guidance from Traditional Owners, and with their approval, share and adopt the seasons used by the many nations in this country. Many of these seasonal systems will challenge Western minds, such as mine, particularly with their fluidity around dates and definition, but that challenge would be a good thing.

As acknowledged upfront in the book, though, there is little incentive for us to change seasons. Maybe it's just curmudgeonly old taxonomists who worry about such things.

Science Victoria, September 2023[29]

Postscript

When asked to contribute this piece on seasons to a special issue of the Royal Society of Victoria's *Science Victoria* magazine focusing on Victoria's flora, I updated the script used for my 2014 ABC RN's *Ockham's Razor* reading.[30] At the end, I reflected on how my own changing views on the seasons.

What hasn't changed is my frustration with the reporting of seasonal changes. In a typical late winter newspaper story from August 2023,[31] my colleagues at Royal Botanic Gardens Victoria confirm yet another 'early spring'. Flowers are blooming early. The warm temperatures are unexpected. Nature is responding to climate change. All true to varying degrees but without acknowledging that the southern Australian spring – as interpreted through familiar plant flowerings – never starts on 1 September.

A decade after finishing my book, I don't know that my system is any better at representing seasonal change than the classic four. Plant flowering (along with fruiting, autumn leaf fall and many other seasonal changes) occurs in overlapping waves, with each species tracking climate and day length in different ways. There are not four, or five, neatly circumscribed seasons and the more nuanced and flexible seasonal systems of the Traditional Owners are a better representation of cyclical change.

The bigger issue is the degree to which climate change will decouple seasonal relationships between different parts of an ecosystem.[32] That is even more complex than trying to define meaningful seasons, and of course much more important.

It is time to ditch Victoria's floral emblem

Common heath an underwhelming representation of a State

With the Victorian election sorted, I feel brave enough to raise the vexing matter of our inadequate State floral emblem.

First of all, can anyone actually name Victoria's floral emblem? For those of you who can't recall, it is the pink common heath. Native to south-east Australia, this is a small shrub with clusters of tube-like flowers that attract honey-eating birds.

I am not alone in thinking the common heath is an underwhelming representation of anything, let alone a whole state. Several concerned citizens – whom I'm too polite to mention here – have encouraged me to find a replacement.

Pink heath has had a good run. In 1958, Victoria became the first Australian State to adopt a floral emblem when Parliament, led by then-Premier Henry Bolte, proclaimed the pink form of the common heath as our totemic plant.

The common heath, *Epacris impressa*, is pretty enough but hardly impressive. *En masse*, in a good flowering year, the common heath will give the local scrubland a soft pink, or sometimes white or red, hue. The plant itself, though, is best described as scrappy. It isn't difficult to grow and is almost impossible to turn into an arresting floral display – in the garden or in a vase.

Line it up beside the waratah from New South Wales or the kangaroo paw from Western Australia, or any of the other emblematic blooms, and Victoria's lacks a certain something.

There is no obvious bold and beautiful plant that sums up Victoria now and its human history over tens of thousands of years. A starting point might be a Victorian endemic, a species that grows only within our State border, but we have few of those. The common heath itself is equally common in Tasmania.

Some of the more recognisable and most loved plants are trees. A few months ago, I was part of an ABC *Catalyst* judging panel that selected the mountain ash, *Eucalyptus regnans*, as Australia's Favourite Tree. This, the world's tallest flowering plant, also grows in Tasmania, but that State is already represented by the Tasmanian blue gum, *Eucalyptus globulus*.

The river red gum, *Eucalyptus camaldulensis*, was the equally worthy winner of the popular poll. It is a beautiful and personable tree, but as the most widespread tree species in Australia it would hardly highlight our south-east corner of the mainland.

A plant of significance to Indigenous Australians would be timely, if perhaps odd, to associate with a colonial concept of statehood. The Wurundjeri take their name from the manna gum (also called white gum) – *wurun* – and the witchetty grub found in the tree (*djeri*). The tree is often part of the Wurundjeri Woi Wurrung smoking ceremony as part of a Welcome to Country.

The manna gum too is widespread, although only in south-eastern Australian states. There are local variants only found in Victoria but these are hardly household names or particularly distinctive. Still, the species itself should be on any short list.

For a plant with a more flamboyant and well-known flower we could look to our heathlands, to species of grevillea, correa or even a native ground orchid such as caladenia. But none would be immediately recognisable or compete, I think, with the waratah or kangaroo paw.

My current favourite is a banksia of some kind. I'd suggest the silver banksia, *Banksia marginata*, a nuggety plant with pale yellow flowers in a bold flowerhead, not a show-off flower like the telopea but something with a deeper, lasting beauty.

No other State has been bold enough to adopt a banksia, so we could seize the opportunity to claim its magnificent flowerhead. Like the manna gum, silver banksia's distribution gathers up part of New South Wales and South Australia, and it extends into Tasmania, but it is widespread in Victoria. It is worth noting that South Australia has the stunning Sturt's desert pea, which occurs in every mainland State except Victoria.

The only problem might be the silver banksia's name. The scientific and English common names honour Joseph Banks, one of the first European botanists to gather plants from Australia and a strong advocate for Australia becoming a penal colony.

The decolonising of plant and animal names in science is a work in progress[33] but we could sidestep this issue by seeking permission to use an Aboriginal name such as *woorike*, which I understand is from the Wurundjeri Woi Wurrung language.

This is the best I can offer, but I'm happy to receive suggestions. Keep in mind, I've been trying for more than a decade to rid Australia of the ill-fitting four seasons it inherited from overseas, and to move Wattle Day a month earlier.

And I do acknowledge there are more pressing issues for the incoming government than its floral emblem. Still, I thought we might all enjoy the distraction.

The Age, December 2022[34]

Postscript

So, how did that go? No death threats but plenty of praise for the common heath. Some liked its subdued and modest nature, arguing that makes it an appropriate emblem for our State (Victoria) and a perfect representative of the Australian bush which is more often muted in colour. Others admired its 'striking colour' and happiness it brings them when in bloom in the Victorian heathland. Its scrappy appearance was also seen as entirely apt by many.

I received a few addition suggestions for replacements, including:

- Myrnong or yam daisy (*Microserias* sp.) with a strong connection to Aboriginal culture.
- Green she-oak (*Allocasuarina paradoxa*), an even less glamorous shrub with barely perceptible flowers.
- Billy buttons (*Craspedia globosa*), with their pert yellow globes of sunshine and long-lasting as a cut flower (you can display it dried).
- Local pigfaces (*Disphyma* and *Carpobrotus*), which are certainly easy to grow but their soft daisy-like flowers might be a little exotic-looking to many.
- Flame grevillea (*Grevillea dimorpha*), a tough and pretty alternative.
- One of the paper daisies (*Xerochrysum*), with pretty, lasting and colourful flowers.

Other banksias were also suggested, such as the old man or saw banksia (*Banksia serrata*).

As for me, I keep returning to the silver banksia or if you like, one of its kin. Let's seize the best Australia has to offer, or if that feels a little ostentatious for Victorians, I think the flame or another grevillea would be a worthy runner up.

Decolonising nomenclature

Planning to rid plant nomenclature of horrific honorifics

You will be familiar with the hostile toppling of statues, particularly those erected by previous generations to celebrate individuals we know today to be violent dictators, extreme colonialists, blatant racists or slave traders.

While you may argue that in some instances interpretative signage might be more effective than removing the offending object, the decision at least to 'do something' is easy for some people. Apart from a few extremists, no-one today would want to honour Adolf Hitler with a statue in the town square.

Similarly with offensive place names. The Wurrundjeri word, *merri-bek*, recently replaced Moreland, the name of a Jamaican slave plantation, as the name for a local government area to the north of the Melbourne CBD. Many creeks in Australia have had their name changed to remove offensive racist terms for Aboriginal people.

Yet due to the arcane (but generally conducive to maintaining universal stability)[35] rules of botanical and zoological nomenclature, the scientific names of some plants and animals honour criminals and universally reviled people – even in one case, Hitler.

These eponyms, as they are called, may honour a person linked to the discovery of the species, such as the collector, a patron or the provider of some critical piece of information leading to its recognition as new to science.

Others may be a chance to show respect to a sovereign or colleague, or a person of some importance or interest to the author of the name. The designation will sometimes reflect a distinguishing characteristic of the organism, such as *darthvaderiana* (acknowledging here that Darth Vader is a fictional person) for a dark-leaved begonia, and *beyonceae* for a horsefly ornamented with 'dense golden hairs'.

An emerging concern to those who name living things, and have responsibility for regulating their naming, is what to do when we now know the person honoured by the name committed crimes or other reprehensible acts. These charges may have been known at the time but considered unexceptional or acceptable, or hidden from those who devised the name.

A debate simmering among taxonomists today is whether we should erase such names from the scientific literature – at least from this point on – and how to do this in an equitable and responsible way.

One concern is over how universal the condemnation should be to action change. For example, the West today might think 'Vladimir Putin' an inappropriate name to apply to any organism but some Russians might disagree. In any country, those on the extreme right or left wing of politics may view certain individuals quite differently.

This is a problem but it isn't insurmountable. The codes of nomenclature for living organisms have mechanisms for deciding whether names should be conserved or rejected on other than purely technical grounds, and they serve us relatively well.[36] While there may not be universal agreement, a strong majority based on a well-argued case would be a good starting point.

Some fear that it might lead to the 'cancelling' of those who were flawed individuals (at least from our perspective) but who overall still made major positive contributions to society or science. One who is sometimes cited is Sir Joseph Banks, after which the genus *Banksia* was named.

Banks was aboard *H.M. Bark Endeavour* when it landed at Botany Bay in 1770 and the botanical collections he made with his colleague Daniel Solander have resulted in Western scientists making thousands of new botanical discoveries and publishing hundreds of scientific works. Banks was also an active patron and advocate for science for much of his life after returning to England.[37]

He was very much part of the colonial project. The plant material he and Solander gathered from Australia was assumed to be theirs for exploitation – whether that was for commerce or science – with no regard for the rights or benefit of the inhabitants of the land from where the plants came. Banks himself showed little empathy with, or understanding of, Indigenous Australians, describing them as 'cowardly and apathetic', worse in his mind than the 'violent' Māori and the 'thieving' Tahitians.[38]

Returning to England, he advocated strongly for Australia to become a penal colony for England, leading to the settlement of Australia by Europeans from 1788 and the catastrophic disconnection between the Indigenous peoples and their Country.

So, do we celebrate his botanical legacy or censure his role in the disposition of Australian Aboriginal people?

The default position is that his name remains attached to the plants we know now as banksias (and for any other plant names honouring Banks). Personally, I would retain that connection given the pivotal role Banks and his collections made to botany, but I'd be happy to have his case heard by a 'Nomenclatural Court' of some kind. If the pain he caused was considered by such a process to outweigh the benefit he brought to Western science, then yes, I would accept a decision to change.

As it happens, this example has been cited by those proposing changes to the rules of botanical nomenclature to allow 'in certain highly prescribed circumstances, the rejection of scientific names that commemorate historical individuals who committed egregious crimes against humanity'.[39]

In this proposal, a case would have to be mounted for each offending name to be reviewed by a specially formed committee and then put to the vote at the next International Botanical Congress (the same place where the new process itself must be considered).[40] The committee would also consider names derived from terms considered culturally offensive today.

While the authors of the proposals argue they are modest and that 'the nomenclatural community has the maturity and wit to deal with issues of in appropriate historical names and epithets in a mature and considered fashion', some people are in strong disagreement with any censuring of names.

As for Banks, there is no suggestion that his recognition should be diminished, despite some dubious views and actions. On balance, and when compared with slave traders and their ilk, *Banksia* would almost certainly remain. On the other hand, *Hibbertia*, named to commemorate slave trader and advocate, George Hibbert, would be one of the early test cases.

Scientific names change for all kinds of good or at least rational reasons. While we may be irritated at losing a familiar name such as *Hibbertia*, we surely benefit as a society and a scientific community by not giving any more attention to what we all agree – if we do – is a person not worthy to be honoured by this yellow-flowered Australian shrub.

I would also encourage greater use of local language – both Indigenous and other community languages – to refer to plants alongside their scientific names. As observed earlier in this chapter,[41] the scientifically sanctioned name need only be used in the scientific literature and where precision is essential; everywhere else, anything goes, with the caveat that in the context of common names, without any constraints, it would be perverse to persist with any name that celebrated a criminal and contemptable individual.

April 2024

Giving a rat's about trees

Where a native rat and its bum are of more interest than the trees and plants that sustain its life

Rare plants? Who gives a rat's arse? Unless perhaps that rat is the vulnerable black-footed tree rat in northern Australia, and the plant happens to be its food or lodging. I have no doubt scat analysis – my literal connection to a rat's bum – of many a threatened Australian mammal has helped conserve a rare plant or two.

But there are many more plants to save. Most of us are blind to the fact that plant species – which humans also need, for food, pharmaceuticals, fibre and, let's face it, fun and fascination – are becoming extinct at an alarming rate.

Since the botanists Joseph Banks and Daniel Solander visited Australia 254 years ago, 571 documented plant species have become extinct worldwide. That's twice as many as birds, mammals (including rats) and amphibians combined.

Royal Botanic Gardens Kew, in London, reckon there are another

82,000 or so plant species queued up behind those 571 on a trajectory to extinction if we do nothing to halt their decline.

The causes are many but are primarily habitat destruction, pollution and, increasingly, global heating. Big, gnarly problems.

Luckily, plants offer a solution to their own demise. The UN's Intergovernmental Panel on Climate Change has proposed a relatively simple solution. One- to two-thirds (estimates vary in news reports) of all human-produced carbon in our atmosphere could be stored away for decades if we reforested 1.7 billion hectares of suitable land.

This amount of land is purportedly available if we exclude urban and cropped areas but include grazing pasture. A tree canopy of 0.9 billion hectares would be enough, made up of a few trees in pasture paddocks right through to complete reforestation in other places.

It would take 1.2 trillion tree plantings. At the going rate in the United States of 30 cents a tree, that's a cost to the world of US $300 billion. Which, as a recent debate around tariffs has revealed, is about 60 per cent of the cost of goods imported into the United States from China each year.

If we invested in this strategy, in part at least, we must plant the right trees. They should be species suited to the emerging climate. They should not need toxic chemicals either to thrive or to control their pests and diseases, and apart from a little additional watering to help them settle in, they will need to survive with whatever water falls or seeps around them.[42]

On top of that, I'd recommend a variety of species, to support wildlife and build resilience into the new forests and woodlands. You might favour local species but, as I'm wont to say,[43] once you stray beyond the indigenous flora there is little other than patriotism (which is no bad thing) to recommend a plant from the other side of your country in favour of one from the other side of the world. Whether local, Australian or exotic, we should not be planting environmental weeds:[44] in time these may displace other species, undoing the good work.

Climate change itself will also have to be considered in tree selection. At Royal Botanic Gardens Victoria we created an international alliance of over 500 botanic gardens in pursuit of solutions to the climate crisis for ornamental and amenity plant collections.[45] At the Melbourne Gardens – a storehouse of 11,500 tonnes of carbon, in case you were wondering – each potential new tree species is mapped against climate models, testing whether a species is likely to survive in 2090.

Of course, in a botanic garden you can fuss over every planting, giving it the very best horticultural care. You may recall Australian Prime Minister Bob Hawke – famous for making ambitious promises – committed in 1989 to planting a billion trees in Australia. Some 700,000 were planted by the time his government lost office 7 years later, but it is unclear how many of those are alive in 2023.

Even a trillion dead saplings will not address the climate crisis. After planting the right trees in the right place, we then need to care for them as if our lives – and that of rats and other creatures – depend on them. Which of course they do.

All this would take a major change of attitude towards trees, and all plant life. To value them intrinsically, or, if you must, for their services to humans, not just for what fruit and hollows they provide to black-footed tree rats. We need to care about, and for plants, before we sow the first seed.

Unpublished, August 2023

Postscript

I wrote this piece in response to yet another breathless news story about something with furry ears, coloured feathers or, at the other extreme, fangs dripping with poison. However, I do think the concept of 'plant blindness' has had its day[46] and I'm glad in some respects it wasn't published by the newspaper I sent it to.

A better way to promote the value of plants and their conservation is to highlight charismatic plants like Australia's dinosaur tree, the

Wollemi pine (*Wollemia nobilis*), or our freaky orchid flowers that are pollinated by randy male wasps. Many animals also don't get the attention they deserve and some get more than they should. Let's accept the same with plants.

Botany on road to extinction

> *We sometimes forget that blue-green algae ruled the Earth for 3 billion years*

I was at a conference last week[47] where it was argued that we are living in a new geological period, the Age of Modern Man, better known as the Anthropocene. This isn't necessarily a good thing. As our impact on the planet grows and climate change starts to bite, the Anthropocene may be one of the shortest geological periods on record. And it looks like we'll take a lot of the planet's plants and animals with us.

The title of my presentation was 'Curing plant blindness and illiteracy'.[48] I spoke about the importance of plants to surviving and prolonging the Anthropocene, and what botanic gardens and botanists can do to help mankind in this time of trouble – from raising the standard of botanical literacy through to investing in seed banks and research.

Returning to my office at Royal Botanic Gardens Victoria I was greeted with news of a more imminent mass extinction, that of the word 'botany' in our universities. Soon there may be no school or department of botany in an Australian university, and few anywhere in the world.

This doesn't mean plants won't feature in the university curriculum and research. New departments of biological science are forming from an amalgam of botanical and zoological schools, sometimes gathering up agriculture and various environmental units. Botany, or plant science as we like to call it these days, will be part an integrated programme of life science.

While it could be argued that these new arrangements simply reflect our better understanding of the world, where plants and animals and various other organisms all interact and interconnect, we run the risk of losing something very fundamental and important: the ability to discern and understand the organisms with which we share our planet.

Some of these new departments may research, study and teach botany as well, or better, than those who used to have the moniker nailed to the front door. While I value history, and the word 'botany' has strong links to my undergraduate and postgraduate years, it isn't for those reasons I question the loss of this botanical identity.

As with school curricula, we keep adding to the university syllabus as new knowledge is created, without any deletion. Molecular biology is immensely important and influential but it has been shoe-horned into biological teaching, taking much of the space previously allocated to understanding the plants (and animals) that carry the molecules. Embracing molecular methods is essential but not at the expense of basic botanical knowledge.

The mega biological departments being created today are generally split into the mode of study – for example molecular biology, environmental biology, evolutionary biology, ecology – which makes some sense but I wonder if in the end we lose too much in the translation. I find cross-disciplinary conferences such as the one last week fascinating and informative, and I welcome any chance to break out of my organisational silo, but you don't need to toss out the baby to bathe in that water.

In the case of the School of Botany at the University of Melbourne, the one I know best, I am hoping the writing remains on the bronze plaque and not on the wall. Its branding is strong. Internationally the School of Botany has a reputation for excellence in research and teaching, and locally it has used its Foundation to raise money for projects such as $1 million for a joint post-doctoral position with Royal Botanic Gardens Victoria, whose own Foundation raised matching funds.

This is because botany is important. Plants and their botanical relatives are survivors. We sometimes forget that blue-green algae ruled the Earth for 3 billion years. It was a long time ago, of course, before the internet and even before the dinosaurs. The dinosaurs themselves stomped around the planet for 160 million years or so before a giant meteorite hit Earth.

We humans have been here for less than half a million years, with close relatives going back 2 million years at most. This is a tiny blip in geological time. Indeed, there is debate around whether we warrant a geological age for ourselves, whether our time on this planet will leave a sufficient mark in what is called the stratigraphy.

Botany has already earned its place in geological time. Close descendants of the aforementioned algae are still alive today, and the flowering plants bloomed for the first time about 140 million years ago. We would do well to understand how they have survived for so long and to apply our modern science in the context of their diversity and biology. I'd like to think the discipline of botany might survive another century or two, or at least to the end of the Anthropocene.

The Australian, March 2014[49]

Postscript

I still debate with myself whether a botany school is what we really need when the evolutionary tree of life is so diverse and complicated, with fungi more closely related to animals and my first love, the algae, all over the tree. Botany in its traditional sense includes green plants, fungi and algae – a cosy grouping but of no biological coherence. Yet those that study them have collaborated well for so long, perhaps it is an arrangement of convenience and mutual benefit. I can't help but conclude, why not?

Let's sing the praises of taxonomists

What species is that? We need to know

As with musicians, painters, poets, philosophers, sportspeople and possibly those seeking further proof of the Higgs boson, taxonomists might seem dispensable.

Taxonomists are people who earn their living from naming and classifying life. With nine of the 11 million or so living organisms on Earth yet to be catalogued they still have some work to do, but in an economically rationalised world this may not be seen as a huge priority.

Similarly, intriguing questions about the origins of life and why our planet is blessed with such a variety of life forms could remain unanswered. Fascinating discoveries about our world, such as the dinosaur tree Wollemi pine, the glow-in-the-dark cockroach and a pygmy sloth living on its own Caribbean Island don't contribute majorly to GDP. One could argue that humans would survive, albeit with diminished lives, if that's all taxonomists did.

On the other hand, having someone who can distinguish between the different mites that infect honeybees has already saved $66 million through better targeted biosecurity and management. Similarly, being able to detect early and accurately potential weed and pest species is worth many millions of dollars to farmers and park managers.

Telling farmers on the Eyre Peninsula which canola varieties they should plant – based on the mix of fungal pathogens found in last year's stubble – depends on accurate taxonomy, and it is saving an estimated $18 million a year in crop losses.

And in recent years, 3,000 different kind of sponge have been discovered on the Great Barrier Reef and nearby, 70 per cent of them new to science. As part of the taxonomic survey of these and other marine creatures over 1,500 new bioactive chemicals have been isolated – any one of which might be the next cure for cancer or obesity.

There is a clear and urgent need for taxonomy, and the taxonomists who do it. Which is why we should all be interested in a debate raging in the scientific arena around whether worldwide our taxonomic effort is increasing or decreasing. In Australia, we know we are losing taxonomists at the rate of two to three per year, and the workforce is ageing.

Worldwide, we are unarguably well shy of the effort needed to catalogue Earth. It has been estimated that to discover and describe all species would take 300,000 taxonomists some 1,200 years, at a cost of more than $350 billion. And this is a conservative estimate.

Do we need to describe everything? With extinction running at 100 to 1,000 times what is called the 'natural rate', you could argue that we are fighting a losing battle and should cut our losses. But if we are to direct our limited resources into conserving the most important parts of our biological diversity, how do we know which species and systems are the most important? Taxonomy.

We don't have to stick to the classical approach with Latin names and pictures, but we do need to map the diversity and be able to compare one area with another. This should concern economists, too:

contemplate the global churn of $100 billion generated by coffee, and the potential loss of 99.7 per cent of Arabica's natural habitat by 2080.

Taxonomists these days spend a lot of time reorganising what they already know, creating new ways to access this information and summarising their knowledge in books, apps and talks. All good and worthy stuff, but meanwhile species continue to go extinct and the ones that survive aren't cataloguing themselves.

With DNA sequencing and increasingly powerful computers, there is now more information available for fine-scale sorting of closely related organisms, but the tough decisions must still to be made by specialists. Yet there are few dedicated positions for taxonomists, and it is hard for them to shine in the open market of science.

For starters, the published work of taxonomists doesn't measure up against their biological colleagues. Not because the quality isn't good or the science strong enough, just because of the way success is calculated.

Taxonomic papers have a comparatively long 'half-life': while a breakthrough medical paper will be cited (listed as a source of information by another scientist) frequently in the first few years following publication and then often barely again thereafter, taxonomic papers are cited infrequently but for hundreds of years after they first appeared. Most publication statistics used for career progression consider only on the first few years since publication.

Then there is the lack of recognition for taxonomic science, even by fellow scientists. And I don't just mean particle physics envy! Taxonomists may gain immortality by having their name appended to a species name (as is the convention in technical papers) but their publications are seldom cited in the reference list of any paper as a source of new information or scientific discovery (both of which are critical parts of species discovery and documentation).

In Australia, we have an immense job ahead of us documenting life on our 7 million square kilometre continent, where more than three-quarters of the native plants and animals are found nowhere else on Earth. In just over 200 years, we have described about a quarter of the estimated half a million or so species.

Are there new sources of food, medicine or building materials out there? Of course. Which species can we afford to lose if life as we know it, and pertinently human life, is to continue? What relatives of crop plants will help us survive climate change? How many and which plants do we need to produce oxygen for us to breathe?

Scientists need to work more collegiately to recognise the contributions of taxonomists. Science bureaucrats, such as myself, need to keep searching for ways to employ more taxonomists.

As a wider community, we all need to support taxonomy and its practitioners – if not to make our life better, then because our survival depends on it.

The Guardian, March 2014[50]

Postscript

This feels like a good place to end this collection of essays. As I said at the start of this chapter, I'm a taxonomist by training *and* inclination. I'm curious about the world, and I like to compartmentalise and to find inherent connections. Taxonomy is also a field of study that is still undervalued and therefore poorly resourced worldwide. There is so much to do.

My suggested title for this opinion piece was 'Higgs boson Blues: the struggle to discover life on Earth'. Largely an excuse to refer to Nick Cave and the Bad Seeds but also emphasising the point that we have plenty to discover here on Earth and that these discoveries can match those of theoretical physics – a touch of that particle physics envy I mention. (And yes, you can add that to the animal envy of the last essay.)

Physicists and astronomers tend to be better at forming a cohesive lobbying force as well as promoting their discoveries to lay audiences. A cute new species – such as the Wollemi pine I mentioned in the last postscript – gets a flash of glory sometimes but it's barely more than a flicker compared to a new celestial object. Maybe the sighting of a new star is short-changed a little, but the apparatus needed to find them is well funded.

That said, citizen science programmes are as active in taxonomy as they are in astronomy, with many of the significant discoveries made by enthusiastic amateurs. We need more of them but we also need the (mostly) professional taxonomists to follow up with the formal testing and publishing of potentially new species.

EPILOGUE

On the last day of March in 2023, I was taking my regular lunchtime walk through Royal Botanic Gardens Melbourne. I tried to do this every day I was in the office, taking a picture or two for posting on Instagram if the light was good and something took my fancy. This day was pleasant enough for walking and photographing, and I took the path around the back of Long Island, through our South China Collection.

As I came to one of the many views in the Gardens across lawn and water, I saw a dozen or so well-dressed folk, mostly men, crossing the bridge on the opposite side of Long Island, near Dallachy Island. They were well spaced but clearly part of a group, each wearing jacket and collared shirt. My initial thought was they were on a break from a conference or workshop nearby.

Then I focused in on the man and woman leading the group. I thought I recognised the man. I wasn't sure but confident enough to change tack and walk briskly – breaking into a trot when behind the bushes shielding me from the group – back the way I'd come. Catching breath just in time, I was ready to meet the group as they turned out of their path and towards where I emerged.

I stepped forward and offered my hand. 'President Obama,' I said, 'I'm Director of these botanic gardens. Welcome!'. We shook hands and he greeted me warmly. The others in his group gathered around, some keeping a respectable (required for security?) distance, and none jumping out to pin me down or check for concealed weapons. The lady next to the President, whom I assumed was his assistant, reached out her hand and said hello too. The three of us chatted for about 10 minutes, before they continued their walk.

Back in the office recovering my composure after this chance encounter, I found a letter on the desk. Unusual these days, as most correspondence is via email. Unusual also because it was from a cat, asking to be an ambassador for the Gardens in lieu of being able to visit. Unlike dogs, cats (and other non-dog pets) were not allowed into the Gardens, even under the control of their owner. Within an hour, I had requested our head of Melbourne Gardens change the regulation allowing the entry of cats on leads (like dogs) and replied to the cat with the good news.

To top off this amazing day, our executive group met later in the afternoon and agreed to return most of Royal Botanic Gardens Melbourne to the riverside vegetation that would have existed there before the arrival of John Batman in 1835 (that's when Melbourne was founded as a settlement). A small 5 hectare portion of 'traditional' botanic garden landscape would be retained around Oak Lawn, but the rest would – over a few years – be cleared, reseeded and managed as a bushland reserve. The cultivation and celebration of exotic plants, including those from interstate, was, we felt, anachronistic. In 2023, nature was best experienced in its authentic form, with local species shaped only by the cultural practices of the Wurundjeri Woi Wurrung and Bunurong peoples.

Not a typical day but noteworthy, I think you'll agree. The sceptical among you may doubt the credibility of all three events. I have no pictures of my encounter with President Obama. I was either too cool, or forgot, to take a selfie – thankfully in hindsight, given I would probably have asked Mrs Obama (who it turned out was the woman next to the President – in my defence, she had done her hair differently to when I'd last seen her on television) to take the picture. As it happens, the President *was* in Melbourne at the time, giving a lecture and attending dinners (where the entry fee was in the thousands of dollars), so his presence in the Gardens was not as unlikely as you may think.

The letter from the cat was an amusing diversion from other administrative matters later that day and I warmed to it. Its owner had taken great effort to assist her charge and my view was 'why not?' While I'm not particularly fond of cats, I am fond of well-reasoned, dispassionate, feasible and wryly amusing propositions. I addressed my reply to the cat and was happy to oblige.

As to the decision to return the botanic gardens to what some have called its 'original state', that was made up; false news if you like. My public answer to such a proposition – and this is on record – begins with a spirited defence of the modern botanic garden as a vital combination of nature, culture and science. It ends with me suggesting that we demolish the National Gallery of Victoria, the State Library of Victoria and the Melbourne Cricket Ground – for starters – before even considering the destruction of a place of even higher cultural and social importance such as a botanic garden.[1] While I would encourage recognition and even partial restoration of natural vegetation within a botanic garden, it trivialises and underestimates all its values to treat its entirety as a dispensable piece of real estate.

Still, as I said at the beginning of this book, being sceptical should not exclude the whimsy and delight to be found in our lives. They can arrive unexpectedly, as they did on this day, or can be sought and created. One must be receptive to the unlikely and the extraordinary or else risk missing something very special. It *was* President Obama, and the cat *did* have a point. On the other hand, what seems perfectly reasonable and worthy at first blush – to return the botanic garden to nature, let's say – may not be wise nor true.

In a salutary piece in 2012, *Guardian* journalist Rebekah Higgitt defines 'scepticism' (from the Greek, *skeptomai*, to think or consider) as 'doubt or incredulity about a particular idea, or a wider view about the impossibility of having certain knowledge'.[2] As Higgitt noted, the latter strays into philosophical scepticism, which needn't trouble us here, but even in general use the term 'sceptical' includes both the questioning of an idea as well as, let's say the *difficulty*, of ever being certain about anything.

In this collection of essays, it will be immediately obvious where I have doubt or incredulity about an idea but also, I hope, that I'm aware we can never have complete certainty about what is true. That doesn't mean we can have varying degrees of confidence and, as good scientists (or scientific thinkers), be ready to discard anything clearly disproven. But my goal, generally, has been to explore a topic with a sensibly open mind and without malice.

Higgitt ends her piece with a reminder that 'the etymology of scepticism implies enquiry and reflection, not dismissiveness'. To dismiss the likelihood of a visiting President would be obstinate. To exclude the possibility of a literary cat would be churlish. Not to at least consider the possibility that a botanic garden is past its use-by date – or that the Moon has some influence on our garden, or that genetically modified organisms might harm the world – would be unscientific and irrational.

After due enquiry and reflection, an idea may be dismissed but, as far as you are able, not those who hold it. For ideas that ring true, a handshake or kind word should be sufficient reward. If, however, your life or someone else's depends on the strength of your conviction – climate change is a good example – there may be need for an iron fist in a velvet glove.

ACKNOWLEDGEMENTS

First, thank you to the CSIRO Publishing team, particularly George Knott for listening to and then championing my idea for this book, Mark Hamilton for guiding me through the writing and publication process, Tracey Kudis for shepherding the book through the editing stages, and Janet Walker for her perceptive and thorough copy-editing. Everyone has been a delight to work with.

I also thank the editors and editorial staff of *Gardening Australia Magazine*, *The Skeptic*, *The Age*, *The Guardian* and *The Australian*. Almost without exception, their interventions have improved my style and lucidity. I acknowledge and thank the publishers of those magazines and newspapers too, for allowing me to reuse essays from these titles.

Sally Heath, now Publisher with Allen & Unwin, gave me the skills and confidence to be a far better writer than I was before we worked together on my memoir, *Evergreen: the Botanical Life of Plant Punk* (published by Thames & Hudson). The editors at CSIRO Publishing continued that education, as do the authors of the fiction and non-fiction I read. Thank you to all.

To my family, particularly Lynda Entwisle, thank you for allowing me the indulgence to write and for the conversations that will have informed some of these essays. My son Jerome Entwisle provided the illustrations, responding directly or sometimes more obtusely to the essay topic. I thank him for finding time around his busy schedule as barrister and parent, and for the clever and charming responses.

ENDNOTES

Website links were correct at the time of printing.

Preface

1 Despite my sharing with him a cartoon by Dog on the Moon from *The Guardian Weekly*, 12 January 2024, which included the line 'Nobody wants to read cartoons about plants'.

2 Walsh NG, Entwisle TJ (Eds) (1997–89) *Flora of Victoria*, vols 2–4. Inkata Press, Port Melbourne.

3 Entwisle TJ, Sonneman J, Lewis SH (1997) *Freshwater Algae in Australia: A Guide to Conspicuous Genera*. Sainty & Associates, Sydney.

4 Entwisle T (2014) *Sprinter and Sprummer: Australia's Changing Seasons*. CSIRO Publishing, Melbourne.

5 Entwisle T (2022) *Evergreen: The Botanical Life of a Plant Punk*. Thames & Hudson, Port Melbourne.

6 Entwisle T (2008–2023) *Talking Plants* (blog). <http://talkingplants.blogspot.com>

7 The series, *Talking Plants*, is no longer available online but individual episodes (e.g. Plant Communication, 26 December 2015 – https://www.abc.net.au/listen/radionational/archived/talkingplants/plant-communication/6967048) - are archived on the ABC website.

8 Entwisle T (2015–2023) Contributions to *Blueprint for Living*, ABC RN. <https://discover.abc.net.au/index.html#/?query=entwisle%20blueprint%20for%20living>

Chapter 1

1 Carroll L (1895) *Alice's Adventures in Wonderland*. Macmillan, London.

2 I return to this topic in Wood-wide web unravels, p. 169.

3 Hallé F (2002) *In Praise of Plants*. (Translated by David Lee) Timber Press, Portland, USA/Cambridge, UK.

4 Pollan M (2013) The Intelligent Plant: Scientists debate a new way of understanding flora. *The New Yorker*, 23 December.

5 Chamovitz D (2012) *What a Plant Knows: A Field Guide to the Senses*. Oneworld Publications, Oxford.

6 See Immortal plants don't live forever, p. 104.

7 For example Entwisle T, Fogarty J (2015) Talking Plants, Summer Series. *ABC RN*, 26 December. <https://www.abc.net.au/listen/radionational/archived/talkingplants/plant-communication/6967048>

8 Again, see Wood-wide web unravels, p. 169.

9 Wohlleben P (2016) *The Hidden Life of Trees: What They Feel, How They Communicate, Discoveries from a Secret World.* (Translated by Jane Billinghurst) Greystone Books, Vancouver/Berkeley.

10 See Blood may flow over the wattle, p. 184, if you are wondering how these relate to Australian acacias.

11 A subject I explore in Gardening in a post-modern world, p. 140, and other essays.

12 To be fair, Wohllenben adds that soil structure and nutrient profile may be less than suitable and that there is unfavourable compaction due to heavy footfall. You might also consider the different climate.

13 Entwisle T (2017) Stupid plants. *The Skeptic* **37**(1), 48–50.

14 McCollom H (2023) Stressed plants are literally crying out for help, new research finds. *Better Homes and Gardens*, 7 April.

15 Entwisle T (2019) Tree music. *ABC RN Blueprint for Living*, 6 April. <https://www.abc.net.au/listen/programs/blueprintforliving/tree-music/10963200>

16 Khait I, Lewin-Epstein O, Sharon R, Sade N, Yovel Y, *et al.* (2023) Sounds emitted by plants under stress are airborne and informative. *Cell* **186**(7), 1328–1336. doi:10.1016/j.cell.2023.03.009

17 The Times of Israel staff (2019) Israeli scientists find a flower they say can hear approaching bees. *The Times of Israel*, 8 January. <https://www.timesofisrael.com/israeli-scientists-say-they-found-flower-that-can-hear-approaching-bees/>

18 Parry W (2012) Holy talking plant! Flower communicates with bats. *Live Science*, 29 July (online). <https://www.livescience.com/15279-bat-flower-echo-acoustics-sonar-leaf.html>. Also Yong E (2015) With sonar-reflecting leaves, plant lures bats to poo in it. *National Geographic*, 9 July (online). <https://www.nationalgeographic.com/science/article/with-sonar-reflecting-leaves-plant-lures-bats-to-poo-in-it>; Simon R, Bakunowski K, Reyes-Vasques AE, Tschapka M, Knörschild M *et al.* (2021) Acoustic traits of bat-pollinated flowers compared to flowers of other pollination syndromes and their echo-based classification using convolutional neural networks. *PLoS Computational Biology*, 16 December. <https://journals.plos.org/ploscompbiol/article?id=10.1371/journal.pcbi.1009706>

19 Entwisle T (2025) Heard it on the grapevine. *Gardening Australia*, February, 62–63.

20 Leitch C (2018) Hands off! Plants don't like to be touched. *Labroots*, 18 December (online). <https://www.labroots.com/trending/genetics-and-genomics/13561/hands-off-plants-don-t-touched>; also Spinks P (2016) Believe it or not: plants respond tenderly when patted or touched. *The Age* 27 May (online). <https://www.smh.com.au/technology/touchy-feely-plants-take-well-to-human-warmth-and-kindness-20160525-gp3bhp.html>).

21 Braam J, Chehab EW (2017) Thigmomorphogenesis. *Current Biology* **27**(11), 863–864. doi:10.1016/j.cub.2017.07.008

22 Chalker-Scott L (2015) The Myth of Stoic Trees. *Washington State University* (website). <https://s3.wp.wsu.edu/uploads/sites/403/2015/03/thigmomorphogenesis.pdf>

23 Entwisle T (2025) Touchy subject. *Gardening Australia*, March, 58–59.

24 Chomley F (2021) 'Nature for Health and Well-being: A Review of the Evidence'. Royal Botanic Gardens Victoria, South Yarra.

25 Wong J (2020) How many house plants do you need to clean the air in a small flat? *New Scientist*, 8 July. <https://www.newscientist.com/article/mg24732900-700-how-many-house-plants-do-you-need-to-clean-the-air-in-a-small-flat/>

26 Entwisle T (2020) A breath of fresh air. *Gardening Australia*, December, 54–55.

27 Entwisle T (2019) Fact or fiction. *Gardening Australia*, November, 48–49.

28 Swinton J, Ochu E, The MSI Turing's Sunflower Consortium (2016) Novel Fibonacci and non-Fibonacci structure in the sunflower: results of a citizen science experiment. *Royal Society Open Science* **3**(5), 160091. doi:10.1098/rsos.160091

29 Elomaa P (2024) Developmental patterning of head-like inflorescences in Asteraceae. *XX International Botanical Congress, Madrid* (response to question at symposium presentation).

30 Majumder PP, Chakravarti A (1976) Variation in the number of ray- and disc-florets in four species of Compositae. *Biology, Environmental Science* (*Semantic Scholar* <https://www.semanticscholar.org/paper/Variation-in-the-number-of-ray-and-disc-florets-in-Majumder-Chakravati/b7957afd468799a5b7a52132730909d73ab81d08>). Note that I say 'it seems' in the text because there is no indication this research was peer-reviewed or accepted for publication.

31 Entwisle T (2019) Curiosity: nature by numbers. *Gardening Australia*, January, 46–48.

32 Entwisle T (2017) How Fibonacci and the golden ratio can make your garden beautiful. *ABC RN Blueprint for Living*, 28 January. <https://www.abc.net.au/news/2017-01-28/fibonacci-golden-ratio-can-make-your-garden-beautiful/8217002>

33 Turner H-A, Humpage M, Kerp H, Herthertington AJ (2023) Leaves and sporangia developed in rare non-Fibonacci spirals in early leafy plants. *Science* **380**, 1188–1192. doi:10.1126/science.adg4014. And reported in *SciTech Daily* (n.d.) <https://scitechdaily.com/natures-secret-code-new-findings-shatter-long-held-beliefs-about-fibonacci-spirals/#:~:text=The%20results%20suggest%20that%20the,have%20another%20type%20of%20spiral>

34 Carl's father devised their surname, Linnaeus, so he could register at university (before then he, like many others of the time, had no need of a surname). The Swedish word 'linn' is used for what we call the linden or lime tree (*Tilia*) and the freshly minted name honoured an impressive specimen of linden growing beside their family home. So it is no wonder Carl had such a passion for plants. Perhaps too much. His classification system for plants was

based largely on their sexual organs, the flowers, and considered by some to be a little too explicit – loathsome harlotry, according to one contemporary commentator.

35 These species are from the ones Linnaeus proposed for his horologium florae, according to Kerner von Marilaun A, Oliver FW, Macdonald MF, Busk MB (1896) *The Natural History of Plants; Their Forms, Growth, Reproduction, and Distribution*. Blackie, London.

36 Vandenbrink JP, Brown EA, Harmer SL, Blackman BK (2014) Turning heads: the biology of solar tracking in sunflower. *Plant Science* **224**, 20–26. doi:10.1016/j.plantsci.2014.04.006

37 Entwisle T (2018) Floral clocks and flowering thyme. *Talking Plants*, 30 January (blog). <http://talkingplants.blogspot.com/2018/01/floral-clocks-and-flowering-thyme.html>. Written to accompany a radio segment on 'Time', for *ABC RN Blueprint for Living*, 27 January 2018. <https://www.abc.net.au/listen/programs/blueprintforliving/last-half-hour-time/9390370>

38 Entwisle T (2022) An everlasting timepiece. *Taking Plants*, 25 January (blog). <http://talkingplants.blogspot.com/2022/01/an-everlasting-timepiece.html>

39 Atamain HS, Creux NM, Brown EA, Gardner AG, Blackman BK, *et al.* (2016) Circadian regulation of sunflower heliotropism, floral orientation, and pollinator visits. *Science* **353**(6299), 587–590. doi:10.1126/science.aaf9793

40 Entwisle T (2018) Hungry leaves. *Gardening Australia*, February, 48–49.

41 Niu J, Liu C, Huang M, Liu K, Yan D (2020) Effects of foliar fertilization: a review of current status and future perspectives. *Journal of Soil Science and Plant Nutrition* **21**, 1–15.

42 Egri A, Horvath A, Kriska G, Horvath G (2010) Optics of sunlit water drops on leaves: conditions under which sunburn is possible. *New Phytologist* **185**(4), 979–987. doi:10.1111/j.1469-8137.2009.03150.x

43 Entwisle T (2018) The burning question. *Gardening Australia*, November, 62–63.

44 Entwisle T (2009) The acid test. *Talking Plants*, 24 June (blog). <http://talkingplants.blogspot.com/2009/06/acid-test.html>, incorporating information from other *Talking Plants* posts on the same subject <http://talkingplants.blogspot.com/search?q=hydrangea+colour>

45 Kodama M, Tanabe Y, Nakayama M (2016) Analyses of coloration-related components in *Hydrangea* sepals causing color variability according to soil conditions. *The Horticultural Journal* **85**(4), 372–379. doi:10.2503/hortj.MI-131

46 Franklin DC (2003) Bamboo and the northern Australian connection. *Flora Malesiana Bulletin* **13**, 275–277; and for more on the topic of when a plant becomes a weed, see First Australian plants, p. 54.

47 Franklin DC (2004) Synchrony and asynchrony: observations and hypotheses for the flowering wave in a long-lived semelparous bamboo. *Journal of Biogeography* **31**(5), 773–736. doi:10.1111/j.1365-2699.2003.01057.x

48 Combining parts of Entwisle T (2020) Boom or bust. *Gardening Australia*, March, 51–52 and Entwisle T (2004) The secret life of plants: bamboozled. *Nature Australia* **28**(3), 72–73.

Chapter 2

1 In his 1994 obituary of fellow American religious leader Ezra Taft Benson (Farewell to a prophet. <https://www.lds.org/ensign/1994/07/farewell-to-a-prophet?lang=eng>), Gordon B. Hinckley noted that Benson understood on the farm 'that without hard work, nothing grows but weeds'. This often quoted saw (I read it most recently on a pop-up flower box on Glenferrie Road in Hawthorn) demonstrates part of the problem. Traditional values encourage us to believe that if it's easy – and weeds are easy – then it must be wrong.

2 Watson D (2014) *The Bush: Travels in the Heart of Australia*. Penguin, Melbourne.

3 CSIRO (2009) Climate change may wake up 'sleeper' weeds. *Phys.org*, 15 April (website). <https://phys.org/news/2009-04-climate-sleeper-weeds.html#:~:text=(PhysOrg.com)%20%2D%2D%20Climate,serious%20damage%20to%20the%20environment>

4 Centre for Invasive Species Solutions (2021) A weedy problem. *Weeds Australia* (website). <https://weeds.org.au/overview/>

5 For more on *Entwisleia* see New species are where you find them, p. 196; for more on Wollemi pine, see Woodford J (2005) *The Wollemi Pine: The Incredible Discovery of a Living Fossil from the Age of the Dinosaurs*. Text Publishing, Melbourne.

6 See I accept extinction, reluctantly, p. 205.

7 According to Royal Botanic Gardens Kew (2017) *The Plant List* (website), there are 304,419. <http://www.theplantlist.org>. The term 'plant' is used here to refer to vascular plants: i.e. angiosperms (flowering plants), conifers and ferns. It does not include bryophytes (mosses etc.), algae or fungi.

8 But see Bad blood, p. 62, and other essays in this chapter.

9 Chapman AD (2009) 'Numbers of Living Species in Australia and the World – Executive Summary' Australian Biological Resources Study, Canberra, ACT. <http://www.environment.gov.au/science/abrs/publications/other/numbers-living-species/executive-summary#plants>. To get the percentage (6.3 per cent), I've used this publication for the total of described and accepted species from Australia, a figure calculated in 2009, while my total for the world is from *The Plant List* (note 7 above), a calculation made in 2016. If I'd used the world figures from this 2009 publication (281,600 for the world, 19,300 for Australia), the percentage would be 6.9 per cent of the world species in Australia.

10 European Climate, Infrastructure and Environment Executive Agency. LIFE Programme (website). <http://ec.europa.eu/environment/life/publications/lifepublications/lifefocus/documents/plants.pdf>

11 A reference to the impacts of the last Ice Age, about 20,000 years ago. I return to this topic in The weed-erness, p. 82.

12 Division of Early Warning and Assessment (2006) 'Africa Environment Outlook 2: our Environment, our wealth'. UN Environment Programme, Nairobi. <https://www.unep.org/resources/report/africa-environment-outlook-2-our-environment-our-wealth>

13 World Conservation Monitoring Centre (1992) *Global Biodiversity: Status of the Earth's Living Resources*. Springer Science & Business Media, Dordrecht.

14 Noting that families vary considerably in size and therefore the information they contain: from thousands of species – such as the orchid family, Orchidaceae, with an estimated 22,000 to 25,000 species – to a single species, as is the case with the Cephalotaceae, including only the Albany pitcher plant (*Cephalotus follicularis*) from Western Australia.

15 All those species arising from a particular ancestor.

16 Orchard AE (1999) Introduction. In *Flora of Australia*. Vol. 1, 2nd edn. (Ed. T Orchard) pp. 1–9. ABRS/CSIRO Australian Biological Resources Study, Canberra; 'Largest angiosperm families in Australia', modified by Wikipedia to AGPII conform with AGPII classification. <https://en.wikipedia.org/wiki/Flora_of_Australia> Note that this table includes only flowering plants, not conifers and ferns, which adds a few more species but doesn't change the percentages significantly.

17 Crisp MD, Cook LG (2013) How was the Australian flora assembled over the last 65 million years? A molecular phylogenetic perspective. *Annual Review of Ecology, Evolution, and Systematics* **44**(1), 303–324. doi:10.1146/annurev-ecolsys-110512-135910

18 Crisp and Cook (2013).

19 Crisp and Cook (2013).

20 As I'll return to later in this chapter.

21 Barker NP, Weston PH, Rutschmann F, Sauquet H (2007) Molecular dating of the 'Gondwanan' plant family Proteaceae is only partially congruent with the timing of the break-up of Gondwana. *Journal of Biogeography* **34**(12), 2012–2027. doi:10.1111/j.1365-2699.2007.01749.x

22 Kundera M (2010) *Encounter*. Faber, London.

23 But see The weed-erness, p. 82.

24 See, e.g. Guiltfree planting, p. 99.

25 Entwisle T (2016) Exotic plants face a prickly reception. *The Sydney Morning Herald*, Sydney, 6 June (online). <https://www.smh.com.au/opinion/exotic-plants-face-a-prickly-reception-on-foreign-soil-20130606-2nt07.html>

26 Chandrasena NR (2023) *The Virtuous Weed: Weedy by Name Only, No Weedy by Nature*. Vivid Publishing, Fremantle.

27 Dwyer J (2023) *Weeding Between the Lines*. Australian Garden History Society, South Yarra.

28 Phillips ML, Murray BR, Pyšek P, Pergl J, Jarošík A, *et al.* (2010) Plant species of the Central European flora as aliens in Australia. *Preslia* **82**, 465–482. <https://www.preslia.cz/article/pdf?id=209>; Dodd AJ, Burgman MA, McCarthy MA, Ainsworth N (2015) The changing patterns of plant naturalization in Australia. *Diversity and Distributions* **21**(9), 1038–1050. doi:10.1111/ddi.12351

29 Royal Botanic Gardens Victoria. *VicFlora*. <https://vicflora.rbg.vic.gov.au>

30 Dodd AJ, Burgman MA, McCarthy MA, Ainsworth N (2015) The changing patterns of plant naturalization in Australia. *Diversity and Distributions* **21**(9), 1038–1050. doi:10.1111/ddi.12351

31 See Boab dreaming, p. 70.

32 Royal Botanic Gardens Victoria. *VicFlora*.

33 Again, see Baobab dreaming, p. 99.

34 Bean T (2007) A new system for determining which plant species are indigenous in Australia. *Australian Systematic Botany* **20**(1), 1–43. doi:10.1071/SB06030

35 Entwisle T (2020) Browned off. *Gardening Australia*, February, 38–40.

36 Entwisle T (2017) In defence of the humble Australian lawn. *ABC News, Opinion, RN* (website). <https://www.abc.net.au/news/2017-03-11/in-defence-of-the-humble-australian-lawn/8335790>

37 Harari YN (2016) *Homo Deus: A Brief History of Tomorrow*. Penguin Books Australia, Docklands.

38 Engels J (2016) Why our lawns are bad for the environment and how to change them for the better. *Permaculture News* (online). <https://www.permaculturenews.org/2016/06/03/why-our-lawns-are-bad-for-the-environment-and-how-to-change-them-for-the-better>

39 'Bad blood is like an egg stain on your chin, you can lick it but it still won't go away' is part of the chorus from Bad Blood, a song on the Bonzo Dog Band album, *Let's Make Up and Be Friendly*. The analogy seems to work well for purple loosestrife, given we may never be sure if we've 'licked it'.

40 Royal Botanic Gardens Victoria. Lythrum. *VicFlora*. <https://vicflora.rbg.vic.gov.au/flora/taxon/42192f56-fd23-4339-9be2-8be596daba5a>; Lythrum salicaria L., *Atlas of Living Australia*. <http://bie.ala.org.au/species/LYTHRUM+SALICARIA>; Lythrum. *Flora of Australia*. <https://profiles.ala.org.au/opus/foa/profile/Lythrum>

41 Australian National Botanic Gardens (based on information provided in 1972) Lythrum salicaria. *Information about Australia's Flora: growing Native plants*. <https://www.anbg.gov.au/gnp/gnp2/lythrum-salicaria.html>

42 Little W, Fowler HW, Coulson J (revised and edited by CT Onions) (1973) *The Shorter Oxford English Dictionary, on Historical Principles*. 3rd edn. Clarendon Press, Oxford.

43 Raver A (2015) Don't bring it here. *Landscape Architecture Magazine*, June, 52–62.

44 Blossey B (2002) Purple Loosestrife. *Ecology and Management of Invasive Plants Program* (website). < https://www.invasive.org/biocontrol/11PurpleLoosestrife.cfm>

45 Edwards KR, Adams MS, Kvet J (1998) Differences between European native and American invasive populations of *Lythrum salicaria*. *Applied Vegetation Science* 1(2), 267–280. doi:10.2307/1478957

46 See First Australian Plants, p. 54.

47 *Atlas of Living Australia*. <http://bie.ala.org.au/species/LYTHRUM+SALICARIA>

48 *Atlas of Living Australia*. <http://bie.ala.org.au/species/LYTHRUM+SALICARIA>

49 Groves RH, Australian Department of Environment and Heritage (Australia), Australian Bureau of Rural Sciences (2003) *Weed Categories for Natural and Agricultural Ecosystem Management*. Bureau of Rural Sciences, Canberra.

50 Centre of Agriculture and Bioscience International (CABI) *Invasive Species Compendium* (website). <https://www.cabidigitallibrary.org/product/QI>

51 Munger GT (2002) Fire Effects Information System: *Lythrum salicaria*.. <https://www.fs.usda.gov/database/feis/plants/forb/lytsal/all.html>

52 South African National Biodiversity Institute (2024). Invasive Alien Plant Alert: *Lythrum salicaria*. <https://www.sanbi.org/resources/infobases/invasive-alien-plant-alert/lythrum-salicaria-1/>

53 Pasiecznik, N (2007). Centre of Agriculture and Bioscience International (CABI)Invasive Species Compendium: *Lythrum salicaria*. <https://www.cabidigitallibrary.org/doi/full/10.1079/cabicompendium.31890>

54 See First Australian Plants, p. 54.

55 Bean (2007).

56 Woolls W (1867) *A Contribution to the Flora of Australia*. F. White, Sydney; J. Ferguson, Paramatta; G. Robertson, Melbourne. <https://archive.org/stream/acontributionto00woolgoog/acontributionto00woolgoog_djvu.txt>

57 Benson D, McDougall L (1997) Ecology of Sydney Plant Species. Part 5. Dicotyledon families Flacourtiaceae to Myrsinaceae. *Cunninghammia* **5**, 330–544. <https://www.botanicgardens.org.au/sites/default/files/2023-09/Volume-5%282%29-1997-Cun5Ben330-544.pdf>

58 Royal Botanic Gardens Victoria. *VicFlora*. <https://vicflora.rbg.vic.gov.au/>

59 Bean T (2007).

60 A common name I prefer to baobab, for reasons I'll explain in a later essay, Boab dreaming, p. 70.

61 See Boab dreaming, p. 70.

62 Pettigrew JD, Bell KL, Bhagwandin A, Grinan E, Jillani N, *et al.* (2012) Morphology, ploidy and molecular phylogenetics reveal a new diploid species from Africa in the baobab genus *Adansonia* (Malvaceae: Bombacoideae). *Taxon* **61**(6), 1240–1250. doi:10.1002/tax.616006

63 Rangan H, Rangan H, Bell KL, Baum DA, Fowler R, *et al.* (2015) New genetic and linguistic analyses show ancient human influence on baobab evolution and distribution in Australia. *PLoS One* **10**(4), e0127582. doi:10.1371/journal.pone.0127582

64 Rangan H (2015) Iconic boab trees trace journeys of ancient Aboriginal people. *The Conversation*, 6 April (online). <https://theconversation.com/iconic-boab-trees-trace-journeys-of-ancient-aboriginal-people-39565>

65 For example, it grows in non-European-modified habitats, it is not invasive, and it has local pests and diseases. Although current populations display limited genetic or other diversity, neither do many other clearly native plant species.

66 See First Australian Plants, p. 54.

67 ABC (2015) Research findings back up Aboriginal legend on origin of Central Australian palm trees. *ABC News*, 3 April (online). <http://www.abc.net.au/news/2015-04-03/aboriginal-legend-palm-tree-origin-central-australia-research/6369832>

68 Nano C (2008) 'National Recovery Plan for the Central Australian Cabbage Palm *Livistona mariae* subsp. *Mariae*'. Department of Natural Resources, Environment, The Arts and Sport, Darwin. <https://www.dcceew.gov.au/environment/biodiversity/threatened/recovery-plans/central-australian-cabbage-palm-livistona-mariae-subsp-mariae-2008>

69 Kondo T, Crisp MD, Linde C, Bowman DMJS, Kawamura K, *et al.* (2012) Not an ancient relic: the endemic Livistona palms of arid central Australia could have been introduced by humans. *Proceedings of the Royal Society*, Series B **279**(1738), 2652–2661. doi:10.1098/rspb.2012.0103

70 See Boab dreaming, p. 70.

71 ABC (2015).

72 See previous two essays, Boab dreaming, p. 70, and Palm gods from the north, p. 75.

73 Rossetto M, Ens EJ, Honings T, Wilson PD, Yap J-YS, *et al.* (2017) From Songlines to genomes: prehistoric assisted migration of a rain forest tree by Australian Aboriginal people. *PLoS ONE* **12**(11), e0186663. doi:10.1371/journal.pone.0186663

74 See Australia should scrap the four seasons, p. 213.

75 Ens EJ, Rossetto M, Costello O (2023) Recognising Indigenous plant-use histories for inclusive biocultural restoration. *Trends in Ecology and Evolution* **38**(10), 896–898. doi:10.1016/j.tree.2023.06.009

76 A topic I cover in some detail in The weed-erness, p. 82.

77 Clode D (2014) Seeing the wood for the trees. *Australian Book Review* **366**, 40–50. <https://www.australianbookreview.com.au/abr-online/archive/2014/125-november-2014-no-366/2215-seeing-the-wood-for-the-trees>

78 Pearce F (2015) *The New Wild: Why Invasive Species Will Be Nature's Salvation*. Beacon Press, Boston.

79 Head L (2014) Living in a weedy future: insights from the garden. In *Rethinking Invasion Ecologies from the Environmental Humanities*. (Eds J Frawley and I McCalman) pp. 87–99. Routledge, Abingdon.

80 Pearce F (2015) Invasive weeds. *Talking Plants ABC RN*, 19 December. <https://www.abc.net.au/listen/radionational/archived/talkingplants/invasive-weeds/6966914>

81 Smyth T (2018) *Napoleon's Australia: The Incredible Story of Bonaparte's Secret Plan to Invade Australia*. Random House, Australia.

82 Rolls EC (1984) *They All Ran Wild: The Animals and Plants That Plague Australia*. Angus & Robertson, Sydney. (Cited by Tout-Smith D (2003) Acclimatisation Society of Victoria. *Museums Victoria Collections* (website). <https://collections.museumsvictoria.com.au/articles/1803>)

83 Dwyer J (2018) Blackberry. *Australian Garden History* **29**(4), 4–7. (Quoting from a recollection by Mrs James Fraser, cited by Tindale B (1959) Baron von Mueller gave us blackberries! *Victorian Naturalist* **76**, 33.)

84 Gillbank L (1986) The origins of the acclimatisation society of Victoria: practical science in the wake of the Gold Rush. *Historical Records of Australian Science* **6**, 359–374. <https://www.publish.csiro.au/hr/pdf/HR9860630359>

85 Gillbank L (2007) Of weeds and other introduced species: Ferdinand Mueller and plant and animal acclimatization in colonial Victoria. *Victorian Naturalist* **124**, 69–78. <https://www.biodiversitylibrary.org/page/51258756#page/5/mode/1up>

86 Spencer RD (2006) Managing weeds in Australian botanical gardens. In *Proceedings of the Fifteenth Australian Weeds Conference*. (Eds C Preston, JH Watts and ND Crossman) pp. 679–682. Weed Management Society of South Australia, Adelaide.

87 Groves RH, Boden R, Lonsdale WM (2005) 'Jumping the Garden Fence: Invasive Garden Plants in Australia and their environmental and agricultural impacts'. WWF-Australia, Ultimo. <https://wildlife.lowecol.com.au/wp-content/uploads/sites/25/Jumping-The-Garden-Fence.pdf>

88 Robin L, Moore J, Willoughby S, Maroske S (2011) Aliens from the Garden. In *Aliens from the Garden*. (Eds C Whitzman and R Rincher) pp. 1–10. State of Australian Cities Research Network, Melbourne. <https://apo.org.au/sites/default/files/resource-files/2011-12/apo-nid59951.pdf >

89 Groves, Boden, Lonsdale (2005).

90 Virtue JG, Spencer RD, Weiss JE, Reichard SE (2008) Australia's Botanic Gardens weed risk assessment procedure. *Plant Protection Quarterly* **23**(4), 166–178. <https://www.researchgate.net/publication/279905700_Australia%27s_Botanic_Gardens_weed_risk_assessment_procedure>

91 Groves, Boden, Lonsdale (2005).

Chapter 3

1 Richardson T (2014) Breathing new life into garden conservation. *Australian Garden History* **26**(2), 3–4. <http://search.informit.com.au/documentSummary;dn=667573277873397;res=IELHSS>

2 Entwisle T (2005) Marx, the revolutionary Brazilian. *Talking Plants*, 5 May (blog). <https://talkingplants.blogspot.com/2015/05/marx-revolutionary-brazilian.html>

3 See Boab dreaming, p. 70.

4 See The multicultural garden, p. 118.

5 Entwisle T (2022) *Evergreen: The Botanical Life of a Plant Punk*. Thames & Hudson, Port Melbourne.

6 This introduction is adapted from notes prepared for the opening of the Australian Landscape Conference in Melbourne, 19 September 2015.

7 Berg N (2015) In the weeds: little-loved plants win the affection of Future Green Studio. *Landscape Architecture Magazine*, September, 62–70.

8 For more on this topic, see the essays in Chapter 2.

9 See The multicultural garden, p. 118.

10 Prescott RTM (1974) *W.R. Guilfoyle 1840–1912: The Master of Landscaping*. Oxford University Press, Melbourne.

11 An expanded version of notes prepared in the first instance for my fortnightly chat with Angela Catterns on ABC Sydney Radio Breakfast, in 2003.

12 Randall R (2007) 'The Introduced Flora of Australia and its weed status'. CRC for Australian Weed Management, Perth. <https://www.iewf.org/intro_flora_australia.pdf>

13 See First Australian plants, p. 54, for some background on what makes a plant 'naturalised'.

14 As Patron of *Plant Trust (Garden Plant Conservation Association of Australia Inc.)*, I can recommend this association for anyone interested in the preservation of plant species in gardens.

15 Hallé F (2002) *In praise of plants*. Timber Press, Portland, OR/Cambridge ,UK.

16 Lynch AJJ, Barnes RW, Cambecèdes J, Vaillancourt RE (1998) Genetic evidence that *Lomatia tasmanica* (Proteaceae) is an ancient clone. *Australian Journal of Botany* **46**(1), 25–33. doi:10.1071/BT96120

17 Entwisle T (2006) The secret life of plants: immortal plants. *Nature Australia* **28**(7), 72–73. <https://museum-publications.australian.museum/media/dd/documents/AMS389_28_07_2005-2006_LowRes.c120745.pdf>

18 Pegg G, Shuey L, Giblin F, Price R, Entwistle P, *et al.* (2021) 'Fire and rust – impact of myrtle rust on post-fire regeneration'. Threatened Species Recovery Hub, Queensland. <https://www.nespthreatenedspecies.edu.au/media/vr4fpfvl/8-3-5-fire-and-rust-interim-report_v4.pdf>; Australian Department of Agriculture, Water and the Environment (2021) 'Conservation Advice for *Uromyrtus australis* (Peach Myrtle)'. Department of Agriculture, Water and the Environment. <http://www.environment.gov.au/biodiversity/threatened/species/pubs/8830-conservation-advice-23112021.pdf>

19 Killicoat P, Puzio E, Stringer R (2022) The economic value of trees in urban areas: estimating the benefits of Adelaide's street trees. *Treenet* (online). <https://treenet.org/resource/the-economic-value-of-trees-in-urban-areas-estimating-the-benefits-of-adelaides-street-trees/>

20 AECOM (2017) 'Green infrastructure: a vital step to brilliant Australian cities'. An AECOM Brillian Cities Report (online). <https://aecom.com/brilliantcityinsights/brilliant-cities-insights-greening/>

21 Carroll C (2024) Valuing our trees. *Botanic News* **autumn**, 7–8. The $56 million value is based on the 'Burnley Tree Valuation Method 2005 (revised)', and the likely higher value if they used the City of Melbourne methodology.

22 Entwisle T (2017) Accident brings home true value of priceless street tree. *HMAA [Horticultural Media Association of Australia] News* **winter**, 16.

23 Entwisle T (2022) *Evergreen: The Botanical Life of a Plant Punk*. Thames & Hudson, Port Melbourne.

24 Entwisle T (2015) 'Professor Entwhistle is a murderer – the telling tale of ten trees, ten years ago', (video of presentation to) *Trees – Natural & Cultural Values*. Australian Garden History Society (Victorian Branch) forum at the State

Library of Victoria, 29 May. <https://www.youtube.com/watch?v=2nR9E3X7TO4>

25 Entwisle T (2015) Commentary: killing a tree makes room for new life. *The Australian*, 29 May . For more detail on the tree removals see the 'Interlude: Ficus' chapter in Entwisle (2022), pp. 147–168.

26 Taking into consideration my essay on Guiltfree planting, p. 99.

27 This is a word I've always wanted to use in a piece for a general audience, even though it seems little different to 'mixture'. Biologists use the term 'genetic admixture' when previously isolated populations of some species or lineage find themselves together again, and exchange genetic material.

28 See Guiltfree planting, p. 99.

29 Entwisle T (2002) The multicultural garden. *The Gardens* **53**(winter).Magazine of the Friends of Royal Botanic Gardens Sydney.

30 See Patriotism in the Victorian garden, p. 122.

31 Pitcher F (1910) Victorian vegetation in the Melbourne Botanic Gardens. *Victorian Naturalist* **26**, 164–177.

32 See The multicultural garden, p. 118, for further discussion about risks of growing local plants.

33 See again The multicultural garden, p. 118.

34 In Royal Botanic Gardens Victoria (2023) 'Living Collections Plan for Royal Botanic Gardens Victoria' Royal Botanic Gardens Victoria (<https://www.rbg.vic.gov.au/media/i25hn5wl/rbg251-living-collections-plan-15-06-23.pdf>), these beds are flagged for conversion to Australian Rare and Threatened species.

35 In fact, in 2018, I estimated it to be 99 per cent of the 189 countries recognised then by the United Nations. See Entwisle T (2018) 99% visible: nearly all the world in a botanic garden. *Talking Plants* (blog), 4 September 2018. <http://talkingplants.blogspot.com/2018/09/99-visible-nearly-all-world-in-botanic.html>

36 See current version in Royal Botanic Gardens Victoria 'Bioregions', *VicFlora*. <https://vicflora.rbg.vic.gov.au/pages/bioregions>

37 Keating PK (2007) Interview with Lauren Harte. *ABC RN*, 12 July.

38 Entwisle T (2013) Patriotism in the Victorian Garden. *Wildlife Australia* **50**(4), 38–39; a shorter version published as opinion piece: Entwisle T (2015) Melbourne's Separation Tree close to the end of its life. *Sydney Morning Herald*, Sydney, 6 January. <https://www.smh.com.au/opinion/melbournes-separation-tree-close-to-the-end-of-its-life-20150105-12hwch.html>

39 Entwisle T (2023) Guest essay: Vive l'Horticulture de Conservation. *Sibbaldia: The International Journal of Botanic Garden Horticulture* **22**, 1–11. <https://journals.rbge.org.uk/rbgesib/article/view/1997/1903>

40 Spencer R, Cross R (2017) The origins of botanic gardens and their relation to plant science, with special reference to horticultural botany and cultivated plant taxonomy. *Muelleria* **35**, 43–93.

41 Taylor P (2006) *The Oxford Companion to the Garden*. Oxford University Press, Oxford.

42 As documented in this resource prepared for the famous Indian Railway Exam: A Brief Note on Botanical Gardens in India. <https://unacademy.com/content/railway-exam/study-material/general-awareness/a-brief-note-on-botanical-gardens-in-india/>

43 Howard RA (1963) The International Association of Botanic Gardens. *Taxon* **12**, 247–249. <https://onlinelibrary.wiley.com/doi/epdf/10.1002/j.1996-8175.1963.tb00040.x>

44 Heywood V (1969) 'The Botanic Gardens Conservation Strategy'. IUCN-Botanic Gardens Conservation Secretariat; Botanic Gardens Conservation International. <https://portals.iucn.org/library/node/6071>

45 Wyse Jackson PS, Sutherland LA (2000) 'International Agenda in Plant Conservation'. Botanic Gardens Conservation International. <https://www.bgci.org/files/All/Key_Publications/interagendaeng2580.pdf>

46 Entwisle T (2018) The top 10 'first' botanic gardens. *Talking Plants*, 13 March (blog). <http://talkingplants.blogspot.com/2018/03/the-top-10-first-botanic-gardens-plant.html>

47 Howard RA (1954) A history of the botanic garden of St Vincent, British West Indies. *Geographical Review* **44**(3), 381–393. doi:10.2307/212064

48 Entwisle (2022) *Evergreen: The Botanical Life of a Plant Punk.*

49 Slatyer P (2023) Editorial from Hobart conference convener. *Australian Garden History* 34, 2.

50 Spencer R (2019) What is a Garden? *Plants, People, Planet: An Australian Perspective* (website), 1 March 2019. <https://plantspeopleplanet.au/what-is-a-garden/>

51 Johnson S (1785) *Dictionary of the English Language*. 6th edn. JF & C Rivington *et al.*, London.

52 See Top five 'first' botanic gardens, p. 127.

53 Entwisle T (2024) For pleasure and ornament (manuscript). Circulated in April 2024 to members of the Australian Garden History Society as a background paper to considering a change in scope and name for the society.

54 See For pleasure and ornament, p. 135.

55 Entwisle T (2024) Gardening for the past and present. *The Museum (National Museum of Australia)* **23**, 60–61.

56 Raxworthy J (2018) *Overgrown: Practices between Landscape Architecture and Gardening.* The MIT Press, Cambridge, MA

Chapter 4

1 For differing perspectives on this obligation, see these two essays in the previous chapter: For pleasure and ornament, p. 135, and Gardening in a post-modern world, p. 140.

2 Sherrington G (2006) Can kelp work? *The Skeptic* **26**(3), 47–51.

3 Conley D (2003) Agricultural alternatives. *The Skeptic* **23**(2), 20–23.

4 Khan W, Rayirath UP, Subramanian S, Jithesh MN, Rayorath P, *et al.* (2009) Seaweed extracts as biostimulants of plant growth and development. *Journal of Plant Growth and Regulation* **28**(4), 386–399. doi:10.1007/s00344-009-9103-x

5 Craigie J (2011) Seaweed extract stimuli in plant science and agriculture. *Journal of Applied Phycology* **23**(3), 371–393. doi:10.1007/s10811-010-9560-4

6 I have similar concerns about creating seaweed farms to generate biofuels and other products. We have to be careful not to treat the oceans as we have land in pursuit of food and fibre.

7 Entwisle T (2017) Weed is good. *The Skeptic* **37**(2), 32–33.

8 Conley D (2017) Letters to the editor: seaweed woo? *The Skeptic* **37**(3), 61–62.

9 Food and Agriculture Organization of the United Nations. *What Is Organic Agriculture*. Food and Agriculture Organization. <https://www.fao.org/organicag/oa-faq/oa-faq1/en/>

10 Australian Department of Agriculture, Fisheries and Forestry (2022) 'National Standard for Organic and Bio-dynamic Produce. Edition 3.8'. (Last updated November 2022.) Department of Agriculture, Fisheries and Forestry, Canberra. <https://www.agriculture.gov.au/biosecurity-trade/export/controlled-goods/organic-bio-dynamic/national-standard>

11 See Converted to seaweed, p. 146.

12 Conley D (2003) Agricultural alternatives. *The Skeptic* **23**(2), 20–23.

13 Le Page M (2016) Stop buying organic food if you really want to save the planet. *New Scientist*, 30 November.

14 Masterson A (2016) How not to save the world: five wrong 'right things to do'. *Sydney Morning Herald*, Sydney, 15 December (online). <https://www.smh.com.au/world/how-not-to-save-the-world-five-wrong-right-things-to-do-20161213-gta4h4.html>

15 Monbiot G (2022) *Regenesis: How to Feed the World Without Devouring the Planet Paperback*. Allen Lane, London.

16 Naeve L, Lewis D (1989) Pros and cons of organic gardening. *The Iowa Horticulturist* **5**(4), 4–5. <https://dr.lib.iastate.edu/server/api/core/bitstreams/cf61f647-eda1-4cb3-9412-574f09173589/content>

17 Naeve and Lewis (1989).

18 Australian Department of Agriculture, Fisheries and Forestry (2022) 'National Standard for Organic and Bio-dynamic Produce. Edition 3.8'. (Last updated November 2022.) Department of Agriculture, Fisheries and Forestry, Canberra. <https://www.agriculture.gov.au/biosecurity-trade/export/controlled-goods/organic-bio-dynamic/national-standard>

19 Naeve and Lewis (1989).

20 Naeve and Lewis (1989).

21 Morley K, Finch S, Collier RH (2005) Companion planting – behaviour of the cabbage root fly on host plants and non-host plants. *Entomologia Experimentalis et Applicata* **117**(1), 15–25. doi:10.1111/j.1570-7458.2005.00325.x

22 Ofori K, Gamedoagbao DK (2005) Yield of scarlet eggplant (*Solanum aethiopicum* L.) as influenced by planting date of companion cowpea. *Scientia Horticulturae* **105**(3), 305–312. doi:10.1016/j.scienta.2005.02.003

23 See The weed-erness, p. 82.

24 Based on segment 2. Companion planting from Entwisle T (2019) Curiosities: fact or fallacy. *Gardening Australia*, November, 49–50.

25 Sethb (2022) Can you put dead animals in the compost? Yes, and here's how. *That Backyard*, 25 January (blog). <https://thatbackyard.com/can-i-compost-dead-animals/>

26 Klink R van, Laar-Wiersma J van, Vorst O, Smit C (2020) Rewilding with large herbivores: positive direct and delayed effects of carrion on plant and arthropod communities. *PLoS ONE* **15**(1), e0226946. doi:10.1371/journal.pone.0226946

27 Barton PS, McIntyre S, Evans MJ, Bump JK, Cunningham SA, *et al.* (2016) Substantial long-term effects of carcass addition on soil and plants in a grassy eucalypt woodland. *Ecosphere* **7**(10), e01537. doi:10.1002/ecs2.1537

28 Expanded version of '#5. Over my dead body', from Entwisle (2019).

29 Lu D (2023) Does glyphosate cause cancer? Australia's Roundup case against Monsanto will offer a fresh legal answer. *The Guardian*, 1 October (online). <https://www.theguardian.com/australia-news/2023/oct/01/roundup-class-action-monsanto-cancer-glyphosate>

30 Lynas M (2018) *Seeds of Science: Why We Got It So Wrong on GMOs*. Bloomsbury Sigma, London.

31 For context, the higher category 1 ('carcinogenic to humans') ranges from plutonium and tobacco smoke through to sun exposure and processed meats such as bacon (Lynas (2018)).

32 For example, US National Pesticide Information Center. *Glyphosate: General Fact Sheet*. <http://npic.orst.edu/factsheets/glyphogen.html>

33 Lynas (2018).

34 Lynas (2018).

35 Curl C, Hyland C (2023) Glyphosate, the active ingredient in the weedkiller Roundup, is showing up in pregnant women living near farm fields – that raises health concerns. *The Conversation*, 7 December (online). <https://theconversation.com/glyphosate-the-active-ingredient-in-the-weedkiller-roundup-is-showing-up-in-pregnant-women-living-near-farm-fields-that-raises-health-concerns-213636>

36 Lee J (2024) *McNickle v Huntsman Chemical Company Australia Pty Ltd* (Initial Trial) [2024] FCA 807, *Federal Court of Australia*, 25 July (online). <https://jade.io/article/1084164>

37 Higgins TJ (chair) and National Committee for Plant and Animal Science (2007) 'Statement – Gene technology and GM plants'. Australian Academy of Science (online). <https://www.science.org.au/supporting-science/science-policy/position-statements/gene-tech-plants>

38 Lynas (2018).
39 As cited on p. 27 in Lynas (2018).
40 See Glyphosate, punished for the sins of its inventors?, p. 162.
41 The Royal Society of London (2016) 'How does GM differ from conventional plant breeding?' The Royal Society (online). <https://royalsociety.org/topics-policy/projects/gm-plants/how-does-gm-differ-from-conventional-plant-breeding/>
42 Nakaya R (2018) The Wood Wide Web: how trees secretly talk to and share with each other. *The Kid Should See this*. <https://thekidshouldseethis.com/post/the-wood-wide-web-how-trees-secretly-talk-to-and-share-with-each-other>
43 Nelson F (2023) Does a Vast Network of Fungi Connect Forests? Here's What We Know. *Science Alert*, 14 February (online). <https://www.sciencealert.com/does-a-vast-network-of-fungi-connect-forests-heres-what-we-know>
44 Truong C (2023) 'The wood-wide web – a story too good for its own good?' Presentation for *Fungimap*, 14 September [no permanent recording].
45 Wittgenstein L (1921) *Tractatus Logico-Philosophicus*. Harcourt, Brace & Co., New York.
46 Klein T, Siegwolf RT, Körner C (2016) Belowground carbon trade among tall trees in a temperate forest. *Science* **352**(6283), 342–344. doi:10.1126/science.aad6188
47 Sheldrake M (2020) *Entangled Life: How Fungi Make Our Worlds, Change Our Minds and Shape Our Futures*. Vintage Digital, Kindle Edition. Sheldrake also provides an excellent summary of what we do know about the wood-wide web, and both the limitations and potential of that knowledge.
48 Shueman LJ (2021) Mapping the fungi network that lives beneath the soil. *One Earth*, 1 December (online). <https://www.oneearth.org/mapping-the-fungi-network-that-lives-beneath-the-soil/>
49 Leach N (2023) Hidden fungi absorb over a third of Earth's fossil fuel emissions, new study discovers. *BBC Science Focus*, 6 June (online). <https://www.sciencefocus.com/news/hidden-fungi-absorb-over-a-third-of-earths-fossil-fuel-emissions-new-study-discovers/>
50 Trevors JT (2010) One gram of soil: a microbial biochemical gene library. *Antonie van Leeuwenhoek* **97**, 99–106. doi:10.1007/s10482-009-9397-5
51 See Wood-wide web unravels, p. 169.
52 Monbiot G (2022) *Regenesis: How to Feed the World Without Devouring the Planet Paperback*. Allen Lane, London.
53 Hirsch PR, Mauchline TH (2012) Who's who in the plant root microbiome. *Nature Biotechnology* **30**, 961–962.
54 While algae are usually included within the bailiwick of botany, they are not 'plants'. 'Algae' is a group of convenience, including a range of photosynthetic organisms of simple construct scattered across the tree of life. Some so-called green algae are more closely related to plants, but most are more distantly

related than animals and fungi. Still, historically and pragmatically, those who study algae (phycologists, such as myself) are warmly embraced by most botanists.

55 Hirsch and Mauchline (2012).

56 Books such as Lowenfels J, Lewis W (2006) *Teaming with Microbes: A Gardener's Guide to the Soil Food Web* (Timber Press, Portland; Or see the revised edition, retitled *Teaming with Microbes: An Organic Gardener's Guide to the Soil Food Web*, published in 2010) offer plenty more advice on caring for your rhizosphere.

57 Entwisle T (2022) Lunar legends. *Gardening Australia*, June, 54–55.

58 Anonymous (2023) From the archive: growing seeds by moonlight, and a shower of stars at sea (10 January 1923). *Nature* **613**(7943), 251. <https://www.nature.com/articles/d41586-022-04582-8>

Chapter 5

1 Entwisle T (2014) *Sprinter and Sprummer: Australia's Changing Seasons*. CSIRO Publishing, Collingwood.

2 *Eucalyptus* and the bloodwoods, *Corymbia*. At the time I wrote this piece it was mooted there would be 11. There are now proposals to further divide *Corymbia*, the better to reflect the evolutionary relationships revealed through DNA sequencing.

3 Now called the International Code of Nomenclature for algae, fungi, and plants (ICN).

4 Entwisle T (1989) Blood may flow over the wattle. *The Age*, 22 May.

5 Entwisle T (2010) Eight roadside wattles. *Talking Plants*, 8 July (blog). <http://talkingplants.blogspot.com/2010/07/eight-roadside-wattles.html>

6 Mitchell N, Smith B (2021) The wattle war. *Science Friction, ABC RN*, 6 June. <https://www.abc.net.au/listen/programs/sciencefriction/acacia-name-africa-australia-wattle-war-botany-taxonomy/13372220>; with accompanying web story at Smith B (2021) Australia or Africa? The botanical controversy over who can call their plants 'Acacia'. *ABC News*, 20 June (online). <https://www.abc.net.au/news/science/2021-06-20/acacia-name-debate-botany-taxonomy-africa-australia-plants/100221938>

7 ABC Science (2022) Australia's favourite tree: river red gum wins online poll after three rounds of voting. *ABC Science*, 19 July. <https://www.abc.net.au/news/science/2022-07-29/vote-for-your-favourite-australian-native-tree/101210764>

8 International Dendrology Society (2024). *Trees and Shrubs Online: Platanus × hispanica*. International Dendrology Society. <www.treesandshrubsonline.org/articles/platanus/platanus-x-hispanica/>

9 Diop D (2023) *Beyond the Door of No Return*. (Translated by Sam Taylor) Pushkin Press, London.

10 Baum DA (1995) A systematic revision of *Adansonia* (Bombacaceae). *Annals of the Missouri Botanical Garden* **82**(3), 440–471. doi:10.2307/2399893

11 See Boab dreaming, p. 70.

12 Rangan H, Bell KL, Baum DA, Fowler R, McConvell P, *et al.* (2015) New genetic and linguistic analyses show ancient human influence on baobab evolution and distribution in Australia. *PLoS ONE* **10**(4), e0127582. doi:10.1371/journal.pone.0119758

13 Entwisle T (2024) Common ground. *Gardening Australia*, March, 54–57.

14 Cook C (2022) *Tell Them to Be Quiet and Wait*. Atmosphere Press, Kindle Edition.

15 For example, Bishop D (2014) Why most scientists don't take Susan Greenfield seriously. *BishopBlog*, 26 September (blog). <https://deevybee.blogspot.com/2014/09/why-most-scientists-dont-take-susan.html>

16 Now *X*, and less 'perfect' than it was a decade ago. For now, I still prefer the fleeting grit and humour of platforms such as *Bluesky*, one of *X*'s better successors, to the predominance of syrupy affirmations on *LinkedIn*. *TikTok* is what it is.

17 Sadly still in development, but *VicFlora* (<https://vicflora.rbg.vic.gov.au>) and *HortFlora* (<https://hortflora.rbg.vic.gov.au/>) are two rich online resources for those interested in native and garden plants respectively.

18 McColl G (2014) Botanic Gardens Disney app for children causes concerns over in-app marketing. *The Age*, 21 December. <https://www.theage.com.au/national/victoria/botanic-gardens-disney-app-for-children-causes-concerns-over-inapp-marketing-20141219-12ae3e.html>

19 Magdalena C (2018) *The Plant Messiah: Adventures in Search of the World's Rarest Species*. Penguin, London.

20 As almost happened in January 2020.

21 Vucetich JA, Nelson MP, Bruskotter JT (2017) Conservation triage falls short because conservation is not like emergency medicine. *Frontiers in Ecology and Evolution* **5**, 23 May. doi:10.3389/fevo.2017.00045

22 Incorporating ideas from a speech I gave at the opening of an exhibition 'Ascendent and Descendent' by Royal Botanic Garden Sydney's 2008 Artist-in-Residence, Emma Robertson, and reported in Entwisle T (2019) Going extinct. *Talking Plants*, 25 February (blog). <https://talkingplants.blogspot.com/2009/02/going-extinct.html>

23 Monbiot G (2012) That sleighbell winter? It's all part of climate change denial. *The Guardian*, 3 January (online). <https://www.theguardian.com/commentisfree/2012/jan/02/sleighbell-winter-climate-change-denial>

24 Entwisle T (2012) Is it an early spring? Yes and no. *The Guardian*, 4 January. <https://www.theguardian.com/commentisfree/2012/jan/03/early-spring-kew>

25 Entwisle (2014).

26 A 50- or so year project tracking flowering and other seasonal changes in plants growing in Kew Gardens. At the time, featured on the Royal Botanic Gardens Kew website.

27 Entwisle (2014).

28 See Eager for spring in London, p. 208.

29 Based on Entwisle T (2023) Sprummertime: why Australia should scrap the four seasons. *Science Victoria*, September, 25–26. <https://rsv.org.au/sprummertime/>

30 Entwisle T (2014) Sprinter and sprummer: why Australia should scrap the four seasons. *Ockham's Razor ABC RN* (transcript), 29 August. <https://www.abc.net.au/listen/programs/ockhamsrazor/sprinter-and-sprummer/5705564>

31 Touma R (2023) Blooming magnolias and unseasonable fruit: Australia's warmer winter is making spring come early. *The Guardian*, 22 August. <https://www.theguardian.com/environment/2023/aug/22/orchids-blooming-early-australia-when-do-orchids-bloom-spring-where-to-see>

32 Primack RB, Lee BR, Miler TK (2023) Climate change threatens spring wildflowers by speeding up the time when trees leaf out above them. *The Conversation*, 14 March (online). <https://theconversation.com/climate-change-threatens-spring-wildflowers-by-speeding-up-the-time-when-trees-leaf-out-above-them-200975>

33 See Decolonising nomenclature, p. 222.

34 Entwisle T (2022) It's time to ditch Victoria's flower emblem – here's the alternatives. *The Age*, 15 December. <https://www.theage.com.au/national/victoria/it-s-time-to-ditch-victoria-s-floral-emblem-here-s-what-it-could-be-20221215-p5c6kf.html>

35 See Blood may flow over the wattle, p. 184.

36 Again, see Blood may flow over the wattle, p. 184.

37 Entwisle TJ (2022) The Sir Joseph Banks cabinet: the Botanical Bounty of the Endeavour's Voyage to New Zealand and Australia. *The Johnson Society of Australia Inc. Papers* **18**, 29–49.

38 Moore P (2018) *Endeavour: The Ship and the Attitude That Changed the World*. Random House, Australia.

39 Thiele KR, Smith GF, Figueiredo E, Hammer TA (2022) Taxonomists have an opportunity to rid botanical nomenclature of inappropriate honorifics in a structured and defensible way. *Taxon* **71**, 1151–1154.

40 The XX International Botanical Congress in 2024 approved proposals to change botanical names based on racially offensive words or people, and to establish a committee to consider the ethics of all botanical names.

41 In Blood may flow over the wattle, p. 184, and You say boab, I say that's OK, p. 191.

42 As you'll recall from Guiltfree planting, p. 99, and other essays.

43 See, for example, Patriotism in the Victorian garden, p. 122.

44 Again, see Guiltfree planting, p. 99, but also the essays in Chapter 2.

45 Royal Botanic Gardens Victoria. *Climate Change Alliance of Botanic Gardens*. <https://www.rbg.vic.gov.au/initiatives/climate-change-alliance/>

46 See footnote 48.

47 Encountering the *Anthropocene: The Role of Environmental Humanities and Social Sciences*, Sydney Environment Institute, 25–28 February 2014. <https://www.youtube.com/watch?v=4_BNUUV2sFw>
48 'Curing plant blindness and illiteracy' is a title I wouldn't use today. To quote from my memoir, *Evergreen*: 'Both terms may seem a little misleading but the main point is that plants are often invisible or neglected in favour of more charismatic creatures. There has been some recent pushback on the use of the term "blindness" given it is no longer used for visual impairment in humans, and "illiteracy" might also be inappropriate. In any case, it is probably time we botanists got over our inferiority complex regarding animals. Sometimes we protest too much.'
49 Entwisle T (2014) Botany integrated into program of life science. *The Australian Higher Education section*, 12 March.
50 Entwisle T (2014) Let's sing the praises of taxonomists, who help us make sense of our world. *The Guardian*, 13 March. <https://www.theguardian.com/commentisfree/2014/mar/13/lets-sing-the-praises-of-taxonomists-who-help-us-make-sense-of-our-world>

Epilogue

1 Backhouse M (2023) What 'plant galleries' teach us. *The Age*, 2 June, 24.
2 Higgitt R (2012) Skeptics and scepticism. *The Guardian*, 14 November (online). <https://www.theguardian.com/science/the-h-word/2012/nov/13/history-science>

www.ingramcontent.com/pod-product-compliance
Lightning Source LLC
Chambersburg PA
CBHW020844270326
41928CB00006B/542

* 9 7 8 1 4 8 6 3 1 8 2 1 6 *